Information Operations

Brassey's Issues in Twenty-First Century Warfare

Information Operations

Warfare and the Hard Reality of Soft Power

A textbook produced in conjunction
with the Joint Forces Staff College
and the National Security Agency

EDITED BY LEIGH ARMISTEAD

BRASSEY'S, INC.
Washington, D.C.

Library of Congress Cataloging-in-Publication Data

Information operations : warfare and the hard reality of soft power / edited by Leigh Armistead.— 1st ed.
     p. cm. — (Issues in twenty-first century warfare)
  "A textbook produced in conjunction with the Joint Forces Staff College and the National Security Agency."
  Includes bibliographical references and index.
  ISBN 1-57488-698-3 (alk. paper) — ISBN 1-57488-699-1 (alk. paper)
  1. Information warfare. 2. Information warfare—United States. 3. United States—Military policy. 4. War on Terrorism, 2001– I. Armistead, Leigh, 1962– II. Joint Forces Staff College (U.S.) III. United States. National Security Agency/Central Security Service. IV. Title V. Series.

U163.I52 2004
355.3'43'0973—dc22   2004006372

ISBN 1-57488-698-3

Brassey's, Inc.
22841 Quicksilver Drive
Dulles, Virginia 20166

First Edition

10 9 8 7 6 5 4 3 2

Contents

Abbreviations and Acronyms
vii

Acknowledgments
xv

Foreword
xvii

Introduction
1

CHAPTER 1
Foundations:
The Language of Information Operations
9

CHAPTER 2
Intelligence Support:
Foundations for Conducting IO
49

CHAPTER 3
Information Protection:
The Challenge to Modern Bureaucracies
65

CHAPTER 4
Information Projection:
Shaping the Global Village
111

CHAPTER 5
Related and Supporting Activities:
Organize, Train, and Equip
163

CHAPTER 6

Implementing IO:
Recent Campaigns
189

Conclusion:
What Is the Future of Information Operations?
231

Endnotes
235

About the Contributors
259

Index
265

Abbreviations and Acronyms

ACC	Air Combat Command
ACTD	advanced concept technology demonstration
ADDP	Australian defence doctrine publication
ADF	Australian Defence Force
ADFWC	Australian Defence Force Warfare Centre
ADHQ	Australian Defence Headquarters
ADO	Australian Defence Organization
AFB	Air Force Base
AFCERT	U.S. Air Force CERT
AFIWC	Air Force Information Warfare Center
AO	area of operations
AOR	area of responsibility
ARCO R-CND	Army Command Reserve-Computer Network Defense
ASD/C3I	assistant secretary of defense/command, control, computers, and intelligence
ASD/NI2	assistant secretary of defense for networks and information integration
ASD/SOLIC	assistant secretary of defense/special operations and low-intensity conflict
ASP97	*Australia's Strategic Policy*
BATF	Bureau of Alcohol, Tobacco, and Firearms
BIG	Bougainville Interim Government
BIOSG	Bilateral Information Operations Steering Group
BIOWG	Bilateral Information Operations Working Group
BRA	Bougainville Revolutionary Army
C2W	denial and deception
C3	command, control, and communications
C4	command, control, communications, and computers
CA	civil affairs
CAAP	Critical Asset Assurance Program
CAC	common access card
CC	Combatant Commander
CDC	Center for Disease Control
CentCom	U.S. Central Command
CERT	computer emergency response team
CHMP	Coalition Multi-Level Hexagon Prototype
CI	counterintelligence

CIA	Central Intelligence Agency
CIAO	Critical Infrastructure Assurance Office
CIC	communication information centers
CIMIC	Civil Military Cooperation
CINCXXX	Sample Combatant Commanders
CIO	Chief Information Officer
CIP	Critical Infrastructure Protection
CIPO	Critical Infrastructure Program Office
CIPWG	Critical Infrastructure Protection Working Group
CJCS	Chairman of the Joint Chiefs of Staff
CJCSM	Chairman of the Joint Chiefs of Staff Manual
CJCS MOP 30	Chairman of the Joint Chiefs of Staff Memorandum of Policy 30
CNA	computer network attack
CND	computer network defense
CNE	computer network exploitation
CNNWC	Commander Naval Network Warfare Command
CNO	computer network operations
CONUS	Continental US
COTS	commercial-off-the-shelf
CT	counterterrorism
CTIO	counterterrorism information operations
CWB	collaborative white boarding
CZW	command and control warfare
DAA	Designated Approving Authority
DASD	deputy assistant secretary
DASD S&IO	Deputy Assistant Secretary of Defense for Security and Information Operations
DCI	Director of Central Intelligence
DDIO	deputy director for information operations
DDOS	distributed denial of service
DefCon	Defense Threat Conditions
DHS	Department of Homeland Security
DIA	Defense Intelligence Agency
DIAP	Defense-Wide Information Assurance Program
DIAPSG	Defense Information Assurance Program Steering Group
DiD	Defense in-Depth
DII	defense information infrastructure
DIOC	Defense Information Operations Council
DISA	Defense Information Systems Agency
DISN	Defense Information Systems Network
DITSCAP	Defense Information Technology Certification and Accreditation Process
DMZ	de-militarized zones
DoC	Department of Commerce
DoD	Department of Defense

DoDD	Department of Defense document
DoE	Department of Energy
DoJ	Department of Justice
DoS	Department of State
DS	Decision Superiority
DSB-MID	Defense Science Board on Managing Information Dissemination
DSTO	Defence Science and Technology Organization
EA	electronic attack
EC	electronic combat
EDT	Electronic Disturbance Theater
EP	electronic protection
EPA	Environmental Protection Agency
ES	electronic warfare support
EU	European Union
EuCom	U.S. European Command
EW	electronic warfare
ExComm	executive committee
FBI	Federal Bureau of Investigation
FEDCERT	Federal Computer Emergency Response Team
FEDCIRC	Federal Computer Incident Response Cell
FEMA	Federal Emergency Management Agency
FIRST	Forum of Incident Response and Security Teams
1st IOC (Land)	1st Information Operations Command (Land)
FIWC	Fleet Information Warfare Center
FST	field support team
GIG	Global Information Grid
GNOSC	Global Network Operations Security Center
GOTS	government-off-the-shelf
GPS	Global Positioning System
GSA	General Services Administration
GWOT	global war on terrorism
HDR	humanitarian daily rations
HFAC	human factors analysis center
HHS	Health and Human Services
HQAST	Headquarters Australian Theatre
HumInt	human intelligence
I&IA	infrastructure and information assurance
IA	information assurance
IADS	integrated air defense system
IAP	Information Assurance Policy
IATAC	Information Assurance Technology Analysis Center
IAVA	IA vulnerability alert
IC	intelligence community
ICC	Information Coordination Center

ICG	IPI core group
IDS	intrusion detection system
IITF	Information Infrastructure Task Force
IMI	international military information
IMR	Instrumentation for Materials Research
IO	information operations
IOC (Land)	Information Operations Command (Land)
ION	IO Navigator
IOPS	Information Operations Planning System
IOPT	information operations planning tools
IOS	Information Operations Squadron
IO S&I	information operations strategy and integration
IOSPM	*Information Operations Staff Planning Manual*
IOSS	Interagency OpSec Support Staff
IOTC	Information Operations Technology Center
IPB	intelligence preparation of the battlespace
IPI	international public information
IPIIWG	International Public Information Interagency Working Group
IS	information superiority
ISAC	Information Sharing and Analysis Center
ISR	intelligence, surveillance, and reconnaissance
ISSM	Information System Security Manager
ISSO	Information Systems Security Officer
ITALD	improved TALD
IW	information warfare
J2	These are simply different divisions of the Joint Staff–Intelligence
J3	Operations
J5	Plans
J6	Communications
J6K	Information Assurance
J7	Doctrine
J39	Information Operations
JC2WC	Joint Command and Control Warfare Center
JCIWS	Joint Command, Control, and Information Warfare School
JCMA	Joint ComSec Monitoring Activity
JCS	Joint Chiefs of Staff
JDAM	Joint Direct Attack Munitions
JDISS	Joint Deployable Intelligence Support System
JFACC	Joint Force Air Command and Control
JFC	Joint Force Commanders
JFCom	U.S. Joint Forces Command
JFHQ-IO	Joint Force Headquarters for IO
JFSC	Joint Forces Staff College
JIC	Joint Intelligence Centers

JIOAPP	joint information operations attack planning process
JIOC	Joint Information Operations Center
JISE	Joint Intelligence Support Element
JIVA	joint intelligence virtual architecture
JIWSOC	Joint Information Warfare Staff and Operations Course
JOPES	joint operations planning and execution system
JP	Joint Publication
JSC	Joint Spectrum Center
JSCP	joint strategic capabilities plan
JTF	joint task force
JTF-CND	Joint Task Force-Computer Network Defense
JTF-CNO	Joint Task Force-Computer Network Operations
JV 2010	*Joint Vision 2010*
JV 2020	*Joint Vision 2020*
JWAC	Joint Warfare Analysis Center
JWICS	Joint Worldwide Intelligence Communications System
JWRAC	Joint Web Risk Assessment Cell
KE	knowledge edge
KO	knowledge operations
KW	knowledge warfare
LAN	local area network
LIWA	Land Information Warfare Activity
LOAC	Laws of Armed Conflict
LOCs	Lines of Communication
LT	liaison teams
MARCERT	Marine Corps Computer Emergency Response Team
MARFORCND	Marine Forces Computer Network Defense
MasInt	measurement and signature intelligence
MD	military deception
MIST	military information support team
MOE	measures of effectiveness
MOOTW	military operations other than war
MPP	mission program plan
MT	monitoring teams
NATO	North Atlantic Treaty Organization
NAVCIRT	Naval Computer Incident Response Team
NC	network combat
NCA	National Command Authorities
NCS	National Communications Systems
NDP	National Disclosure Policy
NDU	National Defense University
NEST	New and Emerging Science and Technology
NGO	nongovernmental organizations
NHSA	National Homeland Security Agency

NIAC	National Infrastructure Assurance Council
NIAP	National Information Assurance Partnership
NII	National Information Infrastructure
NIPC	National Infrastructure Protection Center
NIST	National Institute of Standards and Technology
NMJIC	National Military Joint Intelligence Center
NMS	national military strategy
NorthCom	U.S. Northern Command
NOSC	Network Operations and Security Centers
NRO	National Reconnaissance Office
NSA	National Security Agency
NSC	National Security Council
NSIRC	National Security Incident Response Center
NSOC/IPC	National Security Operations Center/Information Protect Cell
NSPD	national security presidential directive
NSS	national security strategy
NSTAC	National Security Telecommunications Advisory Committee
NSTC	National Science and Technology Council
NSTISSC	National Security Telecommunications and Information Systems Security Council
NSTISSP	National Security Telecommunications Information System Security Policy
NTIA	National Telecommunications and Information Assurance
ODS	Operation Desert Storm
OEF	Operation Enduring Freedom
OGC	Office of Global Communications
OIF	Operation Iraqi Freedom
OMB	Office of Management and Budget
OPLAN	Operations Plan
OpSec	operations security
OSD	Office of the Secretary of Defense
OSI	Office of Strategic Influence
OsInt	open source intelligence
OSTP	Office of Science and Technology Policy
PA	public affairs
PaCom	U.S. Pacific Command
PCAST	President's Committee of Advisors on Science and Technology Policy
PCC	Policy Coordinating Committee
PCCIP	Presidential Commission on Critical Infrastructure Protection
PD	physical destruction
PD&PA	Public Diplomacy and Public Affairs
PDD	presidential directive decision
PDS	physical defensive system
PI	public information
PIR	priority intelligence requirements

PKI	public key infrastructure
PMG	Peace Monitoring Group
PNG	Papua New Guinea
PNGDF	Papua New Guinea Defence Force
POG	Psychological Operations Group
PRC	People's Republic of China
PSYOPS	psychological operations
PVO	private voluntary organizations
RAAF	Royal Australian Air Force
RAN	Royal Australian Navy
RFI	requests for information
ROE	rules of engagement
RPNGC	Royal Papua New Guinea Constabulary
RPP	regional program plan
67th IOS	67th Information Operations Squadron
S3	simple secure sign on
SAP	special access programs
SBCCOM	US Army Soldier and Biological Chemical Command
SCI	sensitive compartmented information
SEAD	suppression of enemy air defenses
SigInt	signals intelligence
SLOCs	Sea Lines of Communication
SMLI	stateful multilayer inspection
SOCom	U.S. Special Operations Command
SOFA	status of forces agreement
SOLIC	Special Operations and Low-Intensity Conflict
SouthCom	U.S. Southern Command
SpaceCom	U.S. Space Command
SpecOps	special operations
STO	special technical operations
StratCom	U.S. Strategic Command
SWP	Southwest Pacific
TALD	tactical air-launched decoy
TechInt	technical intelligence
TEP	theater engagement plan
TEU	US Army Technical Escort Unit
TMG	Truce Monitoring Group
TPFDD	Time Phased Force Deployment Data
TransCom	U.S. Transportation Command
TEMPEST	is not an acronym but a Security RQMT
TSCP	theater security cooperation plan
UAV	unmanned aerial vehicle
UCP	unified command plan
UCP '99	Unified Command Plan 1999

USAID	United States Agency for International Development
USAMRID	United States Army Medical Research Institute for Infectious Diseases
USCG	U.S. Coast Guard
USCS	U.S. Customs Service
USD(I)	under secretary of defense for intelligence
USD(P)	under secretary of defense for policy
USG	United States government
USIA	United States Information Agency
USMC	U.S. Marine Corps
VOA	Voice of America
VPN	virtual private network
WMD	Weapons of Mass Destruction
WTC	World Trade Center
Y2K	year 2000

Acknowledgments

Many obligations have accumulated as this project has gone the distance. Therefore, most appropriately the editor and contributors of this book would like to thank a succession of Deans of the Joint Command, Control, Communications, and Information Warfare School at the Joint Forces Staff College, including Captain Thomas F. Keeley (U.S. Navy), Colonel John M. Calvert (U.S. Air Force), and Captain Walt Spearman (U.S. Navy, Ret) for their early support and continued encouragement when enthusiasm sometimes faltered. We also appreciate the efforts by Dr. Fred Giessler (National Defense University, Ret) and Dr. Dan Kuehl (National Defense University) for giving us strategic and broad appreciation of the power of information operations.

Foreword

Dr. Dan Kuehl
Information Resources Management College
National Defense University

As I wrote the original version of this foreword, I focused on the unfolding story and imagery of the 9/11 terrorist attacks on the World Trade Center and the Pentagon, and reflected on the evolving and complex synergy between information and national security. As tens of millions of Americans sat, like myself, glued to our TVs and computers—the fact that a significant percentage of this book's readers watched those events not on TV, but on their desktop computer contains a powerful message about the technological changes we are facing—the realization came to me that literally hundreds of millions of others worldwide were watching the same images simultaneously. This was a powerful reinforcement of what we in the information operations community have been arguing throughout most of the past decade—namely, that the world in which we live and work has become an information fishbowl. In fact, the global information environment has become a battlespace in which the technology of the information age—which is the aspect that we all too frequently focus on—is used to deliver critical and influential content in order to shape perceptions, influence opinions, and control behavior.

It is perhaps merely coincidence that the time gap between the two aircraft crashes into the twin towers of the World Trade Center was sufficiently wide to allow for live TV coverage of the second crash—but I doubt it. As we watched in horrified amazement, the critical eighteen minutes between the two attacks allowed for virtually every available TV camera in New York City to be trained on the twin towers and thus capture the dramatic and terrible imagery live. Ironically, the very day before this tragic event, I told a class at the National Defense University that someday we would see a terrorist act staged and timed to be seen by a live TV audience. Little did I imagine that I would see it enacted so soon, and so close to home.

Thus we come to the importance, relevance, and timeliness of this book. Information operations are playing an increasingly important role in our national security affairs, and as the global war on terrorism (and another new acronym, GWOT, has joined our lexicon) clearly indicates, that role will not be confined solely to the traditional battlefield on which tanks, ships, and planes move and fight. This new battlespace is focused on the "wetware," that is, the "gray matter" of the brain in which opinions are formed and decisions made. The most, perhaps only, effective weapon in this battlespace is information, and the hallmarks of that revolution, such as the transparency of events and the global immediacy of coverage, have only heightened the importance and impact of information operations. The attacks of September 11, 2001, provided a ghastly example of asymmetrical warfare that employed information technology and exploited the speed and reach of global connectivity to deliver content that was described as "shocking" and "staggering," which is tremendously indicative of its emotional and potential political impact.

The attempts by the United States government to combat this type of asymmetric warfare have not always led to success. The time interval between the first draft and final edition of this book also saw the short life of the Office of Strategic Influence. Whether or not creating the "Office" was a wise decision is still a debatable issue, but surely no one doubts that influence can be a strategic tool which nation-states and political groups employ on a daily basis. We see daily evidence of not only the power of influence, but of the new technologies available by which to wield it. Therefore everyone in the national security community, from those traditional members such as the uniformed military and diplomatic personnel to new and uncertain members such as broadcasters and webmasters, need to incorporate the full range of information operations into their plans and future missions. That is why I believe this book is so important, as an effort to educate current and future leaders to the capabilities—and vulnerabilities—inherent in this new era of warfare.

Introduction

This book was written to meet not only a perceived gap in the theoretical construct of international relations in the information age, but also to give an update of the changes with respect to the power of information that have occurred over the last fifteen years. From the fall of the Soviet Union to the accomplishments of the allied coalition in Operation Iraqi Freedom, the ability of the United States to conduct an influence campaign has changed immensely in just a few short years. Often people ask what is different about information from other military weapons or forms of power. The most important concept to remember about information is that it is not a weapon per se; it is a process, a way of thinking about relationships. It is about perception, because information is an enabler, a "source multiplier," a tool that increases one's ability to shape the operational environment. At once a strategy, a campaign, and a process that is supported by traditional military forces, information can do this by using planning tools to synchronize, synergize, and deconflict activities in an overall plan to affect the adversary as well as to enable the horizontal integration of these activities across the whole interagency and coalition environment. In this book, it is our intent to explain not only how important information is to the future of warfare, but also how this warfare area is changing the way that the United States government is organized and how it conducts operations in the information age. The bottom line is that we will be examining the ability of certain key activities to manage influencing and shaping campaigns across the whole political spectrum.

During the cold war, the United States and its allies knew who was the enemy. The Soviet Union and the Warsaw Pact were easily the most recognizable of the "threats" to the free world, but other nations such as China, North Korea, Iran, Iraq, Syria, and Libya were also part of the equation. To use an academic term, the bipolar cold war era was an area of "realist" conflicts, with states as the prime actors and anarchy a central theme. Fast-forward—the former Soviet Union is a shadow of its former self, with a population less than the United States and shrinking. Russia's defense budget is less than 2 percent of the U.S. Department of Defense's, and it cannot deploy a number of its forces because of equipment failures. Likewise, North Korea is starting to open up, Iraq is under occupation by the United States, Iran is lurching toward a transformation, Qadaffi is negotiating with the west, and even China is initiating some democratic processes.

So why, in this post-cold war era, when the great threats to mankind are gone or lessened, is the United States under attack? It is because the enemy has changed. There are still "rogue states" out there that can occupy the politicians and give credence to budget appropriations, but other groups including extremist religious factions have also attacked the United States as well. In the post-bipolar era, most of these nongovernmental organizations (NGOs) or terrorist groups are now operating out from underneath the umbrella of either superpower, and therefore they have much more autonomy. What has happened over the last fifteen years, and especially in the last few, has been an explosion of attacks on networks within the United States by a host of organizations. Some are individuals; others are activists, foreign military units, terrorists, and even nation-states. From an information perspective, a large number of viruses, major military operations around the world, and the tragedy of September 11, 2001, are all recent events that will be mentioned or alluded to later in this text. Each of these incidents in their own way has highlighted the vulnerability of not only the Department of Defense (DoD), but the United States government as well, to these new types of warfare.

What the future holds for the military forces and the national security establishment is unclear; however, there will be many times that the United States will be called upon to engage the multitude of threats and opportunities in this unpredictable age in which we live. Information and the incredible advances in technology have drastically changed the structure of world politics, military strategy, economics, information realm activities, and other familiar restraints that epitomized the cold war era. Thus, the authors truly believe that now is time to awaken to the realities of the information age. This book is not written as another high-tech "doomsday" scenario, but instead it is meant to be an update for the millennium, to identify the threats to national security posed by cyberterrorists, rogue states, foreign militaries, and the enemies within our borders, as well as to showcase the opportunities available from a properly orchestrated information campaign. In addition, we also hope that this publication will illustrate the evolving military doctrine and priorities, which enhance the ability of our government to win the information war, thereby attaining national goals in the information age. For if not, this new millennium may spell a very dangerous and destabilizing period for our country.

The First Battle of a New Kind of War

On September 11, 2001, the "Great Satan," as many Islamic extremists have called the United States, was dealt a staggering blow when followers of terrorist Osama bin Laden crashed hijacked jumbo jets into the Pentagon and the twin towers of the World Trade Center. The death toll ultimately exceeded 3,000 people from over one hundred nations around the world. In essence, the United States lost the first battle in a new kind of war. The "Electronic Pearl Harbor" that disciples of information operations (IO) have warned about for years had not materialized as expected, yet the results were arguably as or more severe than predicted. The economic impact on the United States is still yet to be determined and may prove incalculable. Even worse, the horrific events have left an indelible mark on the psyche of U.S. citizens. The image of "fortress America" that has existed throughout much of the twentieth century has suddenly become meaningless in this new millennium. Therefore, the authors argue that the nation must adjust its national security strategy (NSS) to deal with this new threat.

Try to imagine the same scenario if the airplane crashes had been accompanied by infrastructure attacks against the electrical power industry. Imagine the amplified horror if a major portion of the northeast suffered an electrical power outage four hours after the attacks—such as occurred in August 2003—long enough for news of the attacks to have reached a significant portion of the population. This could have spread general panic and paralysis across the country. Likewise can you imagine phone lines jammed as individuals attempted to contact loved ones in the affected areas for a prolonged period? Combine that with gridlocked highways as panicked civilians in the northeast attempted to flee the major metropolitan areas. The emergency services could become overwhelmed by panicked citizens if the flow of information was not quickly restored. Before September 11, 2001, only an imminent nuclear attack could have evoked such panic. However now it is very different, because the rules have changed, and the United States will probably never be the same again.

Time has passed since these horrific events, and as this book goes to press, there are a number of important lessons that practitioners of IO should take from these tragedies. The first deals with the ability of adversaries, considered by many to be "unsophisticated," to effectively engage in informational warfare against the citizens of the United States. Next, we should carefully consider the difficulty encountered by the United States government (USG) to effectively deal with the adversary's propaganda. Finally, it is worthwhile to watch how the United States has attempted to "shape" its response to these informational attacks.

If the Oklahoma City bombing of 1995 left any doubt, the more recent events should make it clear that a determined adversary, using relatively unsophisticated means, can inflict tremendous harm upon our nation. Some would argue that the attacks of September 11, 2001 were sophisticated in their timing and coordination. Nonetheless, the basic idea of crashing a hijacked aircraft into a large building is a relatively simple means of sending a political message and has been possible for years. And make no mistake about it, the perpetrators' intent was to send a message—namely, that the United States is vulnerable and that there are more terrorist acts to come if the United States persists in its current foreign policy in the Middle East. The message was intended to instill fear and spawn chaos, and to varying degrees it succeeded.

This is psychological operations (PSYOPS) in its purest form. Hitler attempted the same approach with his V-series rockets as he indiscriminately lobbed them at England during the Second World War. But Hitler miscalculated the English people's will and determination. Rather than causing capitulation, Hitler's terror weapons deepened British anger and resolve. Likewise, Osama bin Laden seems to have miscalculated the American people as well. Since those tragic days in 2001, the Bush administration has launched two major military operations in the Middle East, defeating adversarial forces in both Afghanistan (Operation Enduring Freedom) and Iraq (Operation Iraqi Freedom). Not withstanding the fact that both of these nations are still relatively lawless, these actions have drastically changed the American presence in this area. In fact, some analysts have stated that the events of September 11, 2001 actually backfired for the al Qaeda terrorists, because not only did they not drive the American infidels from their Holy Lands, but in fact the U.S. military is now running the governments of two of the larger nations in that region.

The openness and freedom that make life in this country so precious to its citizens also make it vulnerable to informational warfare by its adversaries. We are just beginning to learn how freely the terrorists were

able to move amongst us as they plotted their evil. The very laws that protect our civil liberties make it possible for our adversaries to operate in relative obscurity, right under our noses. The psychological impacts of the
attacks were so severe that there is now serious debate about curbing civil
liberties in exchange for security. From these conversations arose a series
of new laws called the Patriot Acts, which many citizens believe have
severely degraded the Bill of Rights. Likewise, since the attacks, the news
media has increasingly focused on doomsday scenarios of chemical and
biological agent attacks, doing little to calm the growing fears of average
Americans. Though, to an extent, this sense of anxiety has quieted down
with the American populace, still the nation had obviously forgotten
about the necessity for cooperation between military and media during
national emergencies, a practice that was so common during World War II.
It can be easy to blame the news media's lack of discretion today, and we
as a nation have to do better.

The openness of our informational system was clearly demonstrated
shortly after the disasters when the Voice of America (VOA) aired an
interview with Afghanistan's Taliban leader, Mullah Mohammed Omar.
This was reportedly done against the wishes of the State Department.[1] The
interview was filled with venomous attacks against the United States and
Western society in general. The apparent lack of cooperation between
the State Department and VOA is typical of a problem that has plagued
the USG since real-time, global media was born with the advent of the
Cable News Network (CNN). In April 1999, President Clinton signed
Presidential Decision Directive (PDD) 68, entitled *International Public
Information,* or IPI, which was an attempt to gain control over the external messages sent abroad from Washington. PDD-68 directed key agencies
to synchronize the public affairs message sent by the USG to foreign audiences so as to avoid sending confusing messages.[2] However, it is obvious
that the desired synchronization is still lacking. According to a CBS radio
news announcement on October 1, 2002, this incident ultimately resulted
in the replacement of the VOA director. To effectively deal with bin
Laden's propaganda machine, the United States' messages must be crystal
clear and spoken with a single, united voice, as part of an IO campaign.

Finally, we would like to briefly address how the United States has packaged its response to the attacks of September 11th. The President was quick
to declare the attacks an act of war against the United States. The President,
Secretary of State, and Secretary of Defense have repeatedly stressed that the
American public should not expect to see another war like we waged during
the twentieth century. This operation will be a different kind of war, waged

across a broad front and utilizing all elements of national power—political, economic, and military. To win, the United States must successfully integrate political action to gain international support, to employ economic measures to gain international support and neutralize the terrorists, thereby sparing utilizing military power to destroy the ability of the terrorists to sustain themselves. The force that binds these elements of national power together is information, and thus we see how IO will play a key role in waging the fight against global terrorism.

Electronic Disturbance Theatre

"Would you recognize a revolution if you were in it?"

Deep in the jungles of southern Mexico, a rebel leader taps on a notebook computer. He is editing his dissertation that will soon be released to the world with a double click of the mouse. It is December 31, 1994, and Sub-Commandante Marcos has just begun a series of revolts throughout the state of Chiapas in the southern region of Mexico, taking control of several villages in the process. The Mexican Army response was immediate, with twelve days of brutal fighting following the insurrection, yet inexplicably the Mexican government halted its operations short. Although the Mexican Army could have finished the suppression of the Zapatistas, they instead began a series of negotiations that continued until 2001. Why did President Zedillo and his cabinet stop their attacks on the Zapatistas? What factors led to the pause in the fighting that has kept all parties at the bargaining table?

Instead of operating an insurrection by holding rallies and conducting violent acts, the Zapatistas sustained their protest through a series of new and innovative acts of IO. In effect, they dominated the information realm, competing with the Mexican government in a creative information campaign which effectively constrained and manipulated the Mexican government over the last few years in an effort to bring about reform in the Chiapas region. Using nongovernment organizations (NGOs) and the media, Marcos spread the plight of the Chiapas people to activists that pressured President Zedillo and his cabinet. The media coverage forced the Mexican government to halt their suppression of the indigenous peoples of southern Mexico and effectively put any national policy under scrutiny.

More recently in 1998, a small group of activists, known as the Electronic Disturbance Theater (EDT) actively supported the Zapatista movement. These four individuals used computer distributed denial of service (DDOS) attacks to bring down key government servers, thereby bringing media attention to the cause of the Zapatistas. A typical scenario for the EDT was to publicize an attack weeks before the actual event. They used chat rooms, Internet advertisements, and computer conferences to promote their next attack and gain publicity.[3] To increase its effectiveness, the EDT signed up thousands of participants for its DDOS attack, using its self-generated program Floodnet. In April 1998, the group used this application to attack Mexican President Zedillo's website, quickly crashing the server. More attacks continued during the summer of 1998, to include the Mexican Interior Ministry and the Mexican embassy in England, with the largest event planned for September 9, 1998. Bulletins were released in late August and the EDT publicized the impending attack with its Open-Ex exhibit at the Art Festival in Linz, Austria during that time as well. The intended targets were President Zedillo, the Frankfurt Stock Exchange, and the Defense Information Systems Agency (DISA) in the United States.

Because these were publicized "performances," officials at DISA were concerned as to how to thwart these attacks. There were many online inquiries by USG military personnel about the Floodnet program and the EDT during this period, in order to gain knowledge about the purpose of the attack and the nature of the Floodnet applet itself. When the actual attack occurred on September 9, 1998, DISA was ready to defend its network. A system administrator at DISA changed the Perl script on the Floodnet applet, which in effect became an electronic countermeasure effort, and in some eyes, an offensive act. This new applet shut down the web browsers of the users that were supporting the attack by EDT.

Is this scenario fact or fiction? This scenario is indeed true and it is a good example of one type of IO at an unclassified level. In particular, the execution of the DDOS by the EDT radically altered the concept of cyberwar and brought a new term into our lexicon, namely, "Hactivism."[4] These Zapatista sympathizers were true innovators and are recognized by their peers as information warriors extraordinaire. A measure of their success is the amount of space devoted to the Zapatista revolt by the media. What was truly interesting was that it really didn't matter to the EDT whether an attack succeeded or not as long as they received publicity. The Floodnet program was simply a "tool" to get media attention for the Zapatista cause. To date, the insurrection in Chiapas has garnered more media attention than any other insurgent group in Mexico, and in 2001 President Vicente Fox of Mexico officially recognized the Zapatistas and supported a peaceful solution to the Chiapas situation.[5]

The Basis for This Book

As originally written, this book was conceived by former instructors at the Joint Command, Control, and Information Warfare School (JCIWS) of the Joint Forces Staff College (JFSC), in Norfolk, Virginia, as a textbook for their students. As professional IO instructors who conduct the only joint information warfare (IW) course in the United States, these military officers normally teach over five hundred personnel each year in a variety of IO and IW courses dealing with operations in the information age and how influence campaigns have changed the way that warfare is conducted around the world. Thus, this book was developed to meet a perceived gap in the education process of IO students, and it is structured to not only teach the capabilities and related activities of IW, but also give an update of the changes in IO that have occurred over the last fifteen years. From the incredible reaction to the demise of the Soviet Union in 1989 to the continual evolution of U.S. foreign policy in 2004, the use and evolution of IO has changed immensely within an incredibly short time. In essence, then, this textbook traces the history of IO not only from a doctrinal standpoint, but also with regards to organizational changes and educational efforts. Although most of the text deals primarily with the United States, there are also large sections on other nations such as Russia, China, and Australia.

Leigh Armistead
Editor

CHAPTER 1

Foundations:

The Language of Information Operations

"There is a war out there, old friend—a World War. And it's not about whose got the most bullets; it's about who controls the information: what we see and hear, how we work, what we think. It's all about the information."[1]

Cosmo

This book is about power and how the face of power has changed immensely over the last fifteen years. The thesis put forth by the authors is that information, as an element of power, is the most transferable and useful force at all political levels, including the systemic structure of international relations in the post-Cold War era. In an attempt to update the arguments set forth by John Arquilla and David Ronfeldt in their seminal book, *The Emergence of Noopolitik: Toward an American Information Strategy*, we will argue that the use of information is changing the idea of what we look for in the power capabilities within the world political structure.[2] It is evident that informational capability, more than any other component of power, is truly crucial to effective foreign policy in this new era, as shown in Diagram 1-1.

This theory is based on the fact that we now live in the information age—an era of networks and international organizations. Nation-states are losing power to hybrid structures within this interconnected architecture. Access and connectivity, including bandwidth, are the two key pillars of these new organizations. Truth and guarded openness are the approaches used in both the private and government sectors to conduct business. Time zones are becoming more important than borders. This will be an age of small groups using networks to conduct swarming attacks that will force changes in policy.[3] Key features of this era include:

- Wide, open communication links where speed is everything
- Little to no censorship, the individual controls his or her own information flow

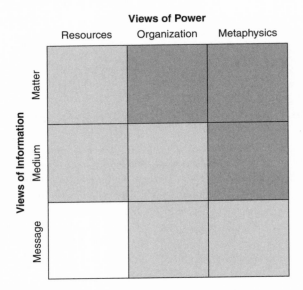

Diagram 1-1   Views of Power and Information

- Truth and quality will surface, but not initially
- Weakening nation-states and strengthening networks

What Is Power?

Power is many things to many different people. Generally people understand its use; they understand who has power and who doesn't. Power is one of those ubiquitous terms that everyone seems to understand but few can actually define. Many scholars, including Morganthau, Dahl, Waltz, Keohane, and Nye, have written works on international relations that have addressed the nature of power and its effects on the global system. Although one could argue about the merits of one definition over another, for the purposes of this book, we will use the following construct: Power is defined as "the ability of A to get B to do something that B would not otherwise do."

Hans Morganthau, in his book *Politics Among Nations: The Struggle for Power and Peace,* defined the elements of national power as geography, natural resources, industrial capacity, military preparedness, population, national character, national morale and the quality of diplomacy and government.[4] Nowhere is the use of information seen as an element of power. This begs the question, have the elements of power changed over the last three decades? If information is now accepted as an element of power, what is different from previous theories? Or, as many believe, has information always been an element of power, it's just that now we have the technology to harness that power? Whatever one believes, the explosion in computer, telecommunications, and media technology has changed our view of power—for better or for worse.

> Traditional measures of military force, gross national product, population, energy, land, and minerals have continued to dominate discussions of the balance of power. These power resources still matter, and American leadership continues to depend on them as well as on the information edge. . . . Information power is also hard to categorize because it cuts across all other military, economic, social, and political power resources, in some cases diminishing their strength, in others multiplying it.[5]

Critics of this new view of power have argued that because less than 20 percent of the world has access to the Internet, that information cannot truly change global politics. This may be true, but the standard has been set, and the benchmark is high.[6] Once people understand the power that is so readily available to them, no longer can dictators rule their countries as fiefdoms. The masses will clamor for the information revolution; as they experience its power, they will threaten the sovereignty of the nation that impedes their progress.

It is evident that ideas about the use and elements of power are changing, as shown in the comparison of Diagrams 1-2 and 1-3. Twenty years ago, Barbara Haskell first discussed the idea of information as power in her article "Access to Society: A Neglected Dimension of Power" in *International Organization*. In 1990, Joseph S. Nye argued for the concept of "soft power," which is described in his book *Bound to Lead*. More recently, in a number of articles starting in the spring of 1996, various authors have highlighted the issues involved with the technological revolution of information. These ideas have also been mirrored by recent books such as *The Rise of the Virtual State* by Richard Rosecrance and *In Athena's Camp* by John Arquilla and David Ronfeldt, both of which discuss the role of information and how it is used to conduct foreign policy.[7]

The idea that information is the most important element of power has not been accepted by all academics. Neorealists still promote ideas of power politics while neoliberals talk about the globalization of the world. Both are correct, but neither camp has been adequately able to explain the changes in world events, especially in the last decade. Other academics realize the power of information, but do not believe that it will change the

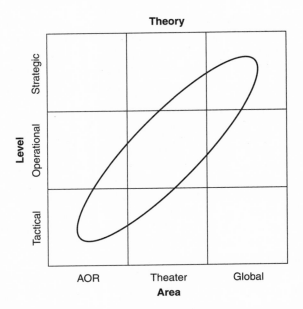

Diagram 1-2   Theoretical Levels of Power

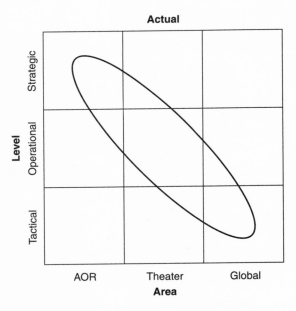

Diagram 1-3   Actual Levels of Power

basic fundamentals of world politics. And there are still a few who are unwilling to realize that they live in a world that is undergoing a revolution. Education and the power of information will play a key role in revitalizing these theories. For, whether these academics realize it or not, changes in technology—especially in the last decade—have rendered their old theories of power obsolete.

> Information technology is the *sine qua non* of both globalization and power—the locomotive on each track. It is integrating the world economy and spreading freedom, while at the same time becoming increasingly crucial to military and other forms of national power. Information technology thus accounts both for power and the process that softens and smoothes power.[8]

There are many factors included by academics in the equation of power. Our belief is that information is now the most important element of power because it is the most transferable. The ability to transfer the power of information is what makes it so useful in the current political situation. Groups, organizations, nation-states and even individuals can now influence policy at the systemic level by using information. This was not necessarily the case a decade ago, but the huge explosion in technology, particularly in telecommunications and media propagation, has vastly changed the power paradigm.

Power in the Cold War Era—What Has Changed?

All of these recent changes have been recognized by a number of individuals from government, military, and academia, as noted in previous sections. And, as mentioned earlier, a number of books and articles have recently recognized how important information is as an element of power. But it is the use of that information and its fungibility that makes it truly different. The ability to transform information, to move it or display its power, all relates directly to its transferability. This is where technology has revolutionized the power structure. The merging of what were once stovepipe and separate areas has opened the access to power for everyone, through the use of information, and has given people a means to distribute it around the world. These ideas are important because they show the true power of information, and that is what has changed.

How one uses information will of course determine whether it is useful or not, but the mere fact that many academics are writing about the power of information shows that something has truly changed. Even the USG has come to realize that, indeed, information is power and has thus begun a process to reorganize itself to take advantage of that fact. This process began with Operation Desert Storm and is continuing today. Lessons learned from that conflict point to the fact that the nation that can control the flow of information is going to win the conflict. Whether that information is in the form of military intelligence, propaganda, electronic wavelengths, or a computer data stream, the ability to manipulate information will be a primary effort of future conflicts.

Military Power and Asymmetric Threats

In a technical sense, military power is often the easiest variable or factor of power to measure. Nation-states have done this since time immortal to compare and contrast military forces. Power throughout the ages has often been ranked solely on the perceived military capability of a nation and the ability of that country to use those forces. This factor is more scientific than some of the other areas, and it has a somewhat useful function of defining weapons and hardware as tools of power. History has, however, generally proven that military capabilities are not so much a reliable factor as many academics would have preferred. For example, how did the United States compare militarily to North Vietnam in 1964? By technical definition, there should have been no contest, yet eleven years later it was American forces that were withdrawing from an ill-fated contest. Likewise, consider events in the former Soviet Union and Afghanistan in 1979. There was a huge disparity in military capabilities, but it was the former Soviet Union who lost that military campaign and returned home vanquished. So why are military forces not a good measure of power? Because, in our view, these weapons and hardware are not fungibile. You cannot adequately translate power in most cases without reverting to total war, which most nations are unwilling to do. Therefore, the most militarily powerful nations are often handicapped in their ability to use their forces to affect desired political outcomes.

Information Operations Theory

From the previous discussion, it should be apparent that the models and theories that have been used by academics to analyze world politics, economics, and military power for the last fifty years are obsolete. Liberalism, realism, and neo-realism are no longer sufficient constructs with which to adequately explain the current dynamics of international power. This can be shown by what is probably the best academic review of IO literature, a Ph.D. dissertation, published in 2002, by Myriam Dunn at the Swiss Federal Institute of Technology in Zurich. Her book, entitled *Information*

Age Conflicts: A Study of the Information Revolution and a Changing Operating Environment, is similar in concept to this text's theory about how IO is radically altering international relations. The prime difference in her approach is Dunn's use of structural realism as her theoretical construct. She confronts the dilemma of the inconsistencies in theory by trying to build a model that delineates key challenges associated with the information age. Dunn also examines all of the traditional international relations theories, and one-by-one dismisses them as inadequate to truly explain the changing environment. Even her proposed choice of structural realism, she admits, has major flaws in its use as a tool for modeling the power of information. In addition, Dunn also recognizes the constraints of all of the different forms of realism that maintain the nation-state as the primary actor. However, in her defense, because a majority of her research was conducted in the 1998–2000 timeframe, she was not able to use the *Noopolitik* book because it had just been published. Perhaps if this theory had been available for Dunn to review, she might have agreed with the basic arguments of this thesis as an alternative to structural realism.

In addition to these changes in academic theory, there has been a substantial change in the nature of strategic, operational, and tactical issues as well. Previous military theories held that strategic concerns were normally a global issue, yet that construct has changed considerably. Now there are numerous events at the tactical level that can quickly elevate to affect the global area of responsibility (AOR) with the use of advanced technology or mass media. Therefore, we propose that, in reality, the new construct for relating the level of military activity cannot be automatically assumed to correlate to a comparable AOR. In fact, as many people realize, with today's new technology often the smallest incidents can spark international or strategic concern, as was shown in the Diagrams 1-2 and 1-3. Likewise, new capabilities that have arisen from the marriage of technology and information have challenged the traditional elements of power including military, diplomatic, and economic factors. These capabilities combined with advanced computing capability and data networking now make options available not only to military and government officials, but also to commercial companies and private citizens that previously did not have such options, as shown in Diagram 1-4. However, the threats to the United States have risen as well.

Attacks on computer systems, negative publicity using the mass media, Internet spamming, and the threat of infrastructure failure have been symptomatic of operations in this new era. No longer is the military and economic might of the United States transferable in many political

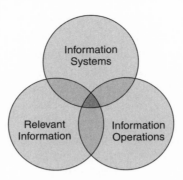

Diagram 1-4   Information Superiority Components

solutions, as in Somalia, for example. General Aideed of Somalia manip-ulated the media to keep the militarily superior U.S. forces off-balance throughout most of the operations during 1993. In fact, with the use of a $600 video camera, Aideed changed forever U.S. foreign policy in the region. It was Aideed, a true information warrior, whose actions in Soma-lia, perhaps more than any other recent U.S. military operation, showed the innate power of information. Although Operation Desert Storm intro-duced the world to the advantages of this revolutionary era, it was in Somalia in October 1993 that the true power of IO came to fruition. By no means is Somalia on par with the United States in a comparison of power of any kind. Yet, because Aideed effectively used the mass media to his advantage, he in fact controlled the flow of events. The use of infor-mation to level the effect of power was instantly recognized and has since been established in doctrine. Since that time, IO has evolved to serve as a model for future asymmetrical conflict and, by implication, international relations. See Diagram 1-5 for a visual representation of IO relationships across time.

IO Theory and Doctrine
Information operations is a formal attempt by the USG to develop a set of doctrinal approaches for its military and diplomatic forces to use and operationalize the power of information. The target of IO is the adversary decision-maker, and therefore the primacy of effort will be to coerce that person, or group of people, into doing or not doing a certain action. U.S. counterterrorism information operations, discussed later, are good examples of the use of this theory in action. To affect the adversary decision-maker, IO attempts to use many different capabilities such as deception, psychological operations, and electronic warfare to shape and influence the information environment.

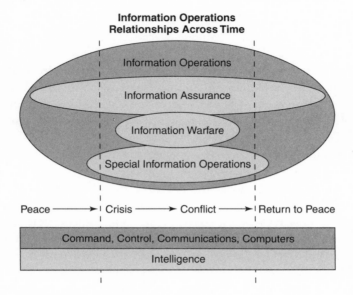

**Information Operations
Relationships Across Time**

Information Operations

Information Assurance

Information Warfare

Special Information Operations

Peace ——→ Crisis ——→ Conflict ——→ Return to Peace

Command, Control, Communications, Computers

Intelligence

Diagram 1-5   IO Relationships Across Time

The capabilities previously mentioned have existed for a long time, but the umbrella term of *IO* is a relatively recent development. Originally developed in 1996 by the government as a component of *Joint Vision 2010* (JV 2010), IO is formally defined as "those actions taken to affect an adversary's information and information systems while defending one's own information and information systems."[9] This official D.O document was written to establish a vision for how the U.S. military will operate in the uncertain future. To implement this vision and achieve "full spectrum dominance," four operational concepts were introduced:

- Dominant maneuver
- Precision engagement
- Full dimensional engagement
- Focused logistics

The essential enabler for all four of these concepts was doctrinally encapsulated as *information superiority*.[10] Defined as "the capability to collect, process, disseminate an uninterrupted flow of information, while exploiting or denying an adversary's ability to do the same," information superiority consists of three components of which information operations is a prime factor.

Despite these developments, IO is still not understood very well. To many people, IO is simply computer warfare. Yet, as discussed earlier in this book, IO is really about much more than that. It is an attempt by the United States to develop a strategy to use all of its capabilities to affect the many issues that it deals with in the post-Cold War era. With these changes in the elements of power has come the realization that, militarily, the United States could not solve all of its problems through kinetic means. IO is therefore an attempt to bring these different facets of power to bear on an adversary in a synergistic manner to achieve our national objectives, as shown in Diagram 1-6.

In June 2000, the United States published *Joint Vision 2020* (JV 2020), the most recent embodiement of a future-oriented military doctrine. This document elevates IO from the conceptualized sub-component level it occupied in JV 2010 to one of two essential elements for success in future military engagements. This latest conceptual document reiterates the dominance of IO within the USG as a key to successful operations over the next two decades. Why is this so? What happened to make this change? Specifically, in the four years between the publication of JV 2010 and JV 2020, much truly has changed within the U.S. military, with many officers and government officials ultimately realizing that future warfare is going to increasingly involve IO. Lessons learned from Rwanda, Bosnia, and Kosovo

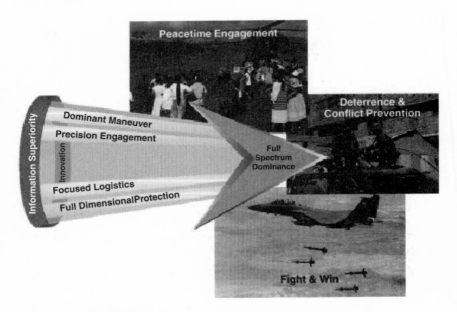

Diagram 1-6 Information Superiority Model

taught the U.S. military the value and inherent power of information. Officials within the government and uniformed services began to understand how effective this new warfare area could be in shaping the battlespace. They witnessed the impact of IO and understood that, if used correctly and early enough in a campaign, IO could even allow one to avoid armed conflict, to not reach the point where the military must be called in to conduct operations.

Differences between Information Warfare and Information Operations

The real key to making IO effective is to ensure that the horizontal integration and coordination of the interagency organizations are conducted early on, that is, in the peacetime environment. As mentioned earlier, IO can be an effective tool for shaping the environment in the pre-hostilities phase, so that the actual need for hostilities may be avoided or minimized. However, this is not always possible. Many military theorists contend that information warfare (IW) is what you use when IO fails. These theorists have recognized one difference, but there are also subtleties between these two warfare areas. The primary doctrinal difference between these two terms is that IW contains six elements and is mostly involved with the conduct of operations during actual combat, whereas IO, on the other hand, includes these six capabilities and sometimes two integrated or related activities. IO is thus broader than IW, and is intended to be conducted as a strategic campaign throughout the full spectrum of conflict from peace to war and back to peace, as shown in Diagram 1-7. Therefore, IO is much more comprehensive than IW, and it is in IO that the full integration across government agencies and with private industry must occur.

A common complaint about IO is that because its definition is so broad, IO is at once everything and it is nothing. The elements, capabilities, and related activities of IW and IO, as listed below, are separate and discrete warfare elements. Most have very old traditions and longstanding histories that do not necessarily mean that every action conducted in these areas is always associated with IO. There are elements of destruction that are not part of an IO campaign, likewise not every public affairs activity has to be tied to information operations. Yet in reality, all elements and their components of national power, in order to succeed, should now be integrated into a satisfactorily planned, designed, and executed information strategy. If this is not done, the United States may not attain its national security goals in the new millennium.

Following is a list of the capabilities and related activities for information operations. These give a foundation for the umbrella theory of IO.

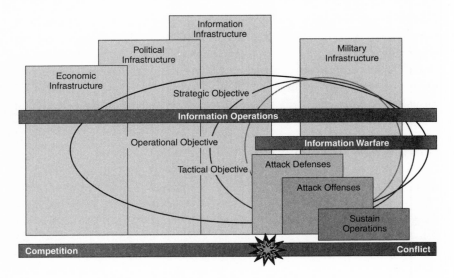

Diagram 1-7   IO versus IW Spheres

Employment of these effects of IO is predicated on the ability of higher headquarters to articulate their intent, direction, restrictions, measures of effectiveness, and timelines for the use of IO capabilities and related activities within their areas of responsibility. Hence, a commander does not derive IO requirements in isolation of theater or strategic requirements. The capabilities and related activities for IO include:

- Civil affairs (CA)
- Computer network attack (CNA)
- Deception
- Destruction
- Electronic warfare (EW)
- Operations security (OpSec)
- Public affairs (PA)
- Psychological operations (PSYOPS)

The concept of IO is thus intended to use these different capabilities and related activities to produce effects in an integrated fashion. Therefore, though one can try to use all eight capabilities and related activities to conduct an operation, more often than not, a good IO plan will probably only incorporate a few of these warfare areas. The basic idea is that one does not always have to resort to kinetic means. Instead, for IO to work properly, the operators must understand the environment, assess their interests and the adversary's pressure points, and then use whichever capability or related activity that will best affect the adversary. IO is thus much more of an intensive study not only of your adversary, but also your own forces, which, perhaps, is more than many current military commanders have grown accustomed to. Yet this idea is not new. Many theorists contend that Sun Tzu (3000 years ago) was the first information warrior; however, the capabilities and related activities of modern information operations have drastically changed since his era.

The Evolution of the IO Doctrine

The development of IO as a major military doctrine in the USG is a relatively new phenomenon, and much of the critical thinking involved began in the early 1980s.[11] The size of the former Soviet Union's military concerned U.S. military analysts and planners. From 1975–1985, the former Soviet Union often outnumbered U.S. conventional forces by three to one; and, though the United States may have had a qualitative advantage, there are times when only sheer numbers count. During this era, the military strategists of the Pentagon were looking for methods to cut down on the former Soviet Union's advantage by attempting to counter traditional strengths with asymmetric non-nuclear attacks. In addition, these analysts noted that the former Soviet Union relied heavily on electronic warfare or *radioelectrionyaborba* (radio electronic combat) in much of its doctrine, and there was a feeling that the USG must combat this threat as well.[12] It was in this era that some of the early ideas about effects-based planning began to evolve.

The demise of the Soviet threat to the continental United States and the shift from bipolar to multi-polar political scenarios has seriously affected the structure of U.S. Force and military doctrine. However, the biggest change in doctrine is due to the huge technological changes that have evolved over the last ten to fifteen years. The advances in computers, software, telecommunications, networks, and so on have revolutionized the way that the USG conducts military operations and have made it the premier armed forces. The magnitude of a series of coalition victories in Operations Desert Storm, Noble Anvil, Enduring Freedom, and Iraqi Freedom clearly showed to the world the overwhelming technological superiority of the U.S. military.

Thus from the lessons learned from these and other experiences since the end of the Cold War, perhaps the most important result has been the rise in the apparent value of information. It has become clear to war-fighters that the side controlling the most information and retaining the ability to accurately manipulate and conduct an influence campaign was going to be victorious. This became apparent immediately after the fall of the Soviet Union, when strategic planners at the Joint Chiefs of Staff began to think about and write new strategy, most of which was highly classified, on the use of information as a war-fighting tool. In fact, the first document, Department of Defense document (DoDD) TS3600.1, was kept at the "Top Secret" level throughout its use due to the restrictive nature of its classification.

Although this publication started a dialogue on IW within the Department of Defense (DoD), its classification ultimately restrained a more general doctrinal exchange. Thus, the need for a strategy to fit these revolutions in technology still existed, so a new concept of command and control warfare (C2W) evolved. Officially released as a Chairman of the Joint Chiefs of Staff Memorandum of Policy 30 (CJCS MOP 30) *Command and Control Warfare* (8 March 1993), this document laid out, for the first time in an unclassified format, the interaction of the different disciplines which gave the war-fighters the IW advantage. C2W as originally defined contained the following five pillars, as shown in Diagram 1-8:

- Destruction
- Deception
- Psychological operations
- Operations security
- Electronic warfare

Intelligence supported these five pillars in order to conduct both offensive and defensive C2W. Likewise, some segments of the military greeted this new concept of warfare with enthusiasm, whereas others were wary of any new doctrinal developments. However, the ability to integrate these different military disciplines to conduct nodal analysis against enemy command and control targets was also highly lauded as a great improvement.

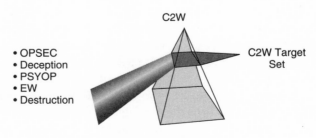

Diagram 1-8   C2W Doctrine

Many units and all four services (Army, Navy, Air Force, Marine Corps) developed C2W cells and began training in this new doctrine throughout the mid-1990s. But there was a conflict between the CJCS MOP 30 and the DoDD TS3600.1 doctrine because IW was a much broader attempt to tackle the issue of information as a force multiplier, whereas C2W was more narrowly defined to apply only to the five pillars. The fact that the United States was writing strategy to conduct operations in peacetime against nations was considered very risky, therefore IW remained highly classified throughout much of the 1990s.

Yet the U.S. military recognized the need to develop commands and agencies to conduct these types of warfare in the information age, and therefore, even though doctrine was still in the formative stage, organizational changes began to occur in the early 1990s. The Joint Electronic Warfare Center at Kelly Air Force Base (AFB) in San Antonio, Texas, was renamed the Joint Command and Control Warfare Center in 1993, and would later be renamed the Joint Information Operations Center (JIOC) in October 1999. The uniformed services also created a number of other new agencies beginning in 1995, including:

- U.S. Air Force—Air Force Information Warfare Center (AFIWC)
- U.S. Army—Land Information Warfare Activity (LIWA)
- U.S. Navy—Fleet Information Warfare Center (FIWC)

In addition to organizational changes by these services, new courses and schools were also being developed to teach new tactics. The National Defense University (NDU) created a School of Information Warfare and Strategy in 1994—a full ten-month-long academic curriculum designed to immerse the National War College students in the academic theory of IW. Held for two years, NDU graduated sixteen students the first year and thirty-two the second; however, the course was subsequently canceled in 1996. This may have been due to a belief that IW instruction needed to be disseminated to a wider audience, so shorter courses and classes were developed instead to teach a larger audience of NDU students. These existed for several years, including a five-day intermediate IW course for mid-grade officers and a two-day IW overview for senior officers, but by mid-2003 all were eventually canceled. The other official DoD joint course on IW is also taught at NDU's Joint Forces Staff College, formerly the Armed Forces Staff College in Norfolk, Virginia. Held for two weeks seven times a year, the Joint Information Warfare Staff and Operations Course (JIWSOC) is aimed primarily at mid-grade officers or civilian equivalent government personnel who are serving in an IO cell or billet with a joint agency.

Doctrine also continued to develop after the publication of CJCS MOP 30. The formation of IW agencies and commands in the 1995 time period not only filled voids in the services but also helped to resolve the conflict in the development of information doctrine and policy within the USG. Thus there was a concerted push for declassification and better understanding of these concepts within the DoD, which resulted in the publication of DoDD S3600.1, *Information Operations* (9 December 1996). By downgrading this document to the "Secret" level, DoD opened IO to a wider audience. In a related effort, the Defense Science Board also published its report on information warfare and defense in November 1996. Together these two documents attempted to clarify the differences from the older doctrine, and for the first time introduced the use of CNA as an IO capability.

Thus, the formation of IW agencies and commands in the 1995–1996 time period also somewhat helped to resolve the conflict in the development of IO doctrine and policy within the USG. However, since DoDD S3600.1 was still classified "Secret," it also limited greater discussion on the differences between IO and IW. Thus the mid-to-late 1990s were also a period of early experimentation. A number of exercises were conducted, elevating the awareness of IO within the military and civilian communities. The CNA operations conducted during the 1996 and 1997 exercises were also particularly effective and drew attention to the fact that the DoD was vulnerable to this type of operation. There were, however, still questions regarding IO definitions and components that would not be fully addressed until the release of the seminal publication, Joint Publication 3-13, *Joint Doctrine for Information Operations* (9 October 1998). For the first time, the DoD released an unclassified document to widely disseminate the doctrinal principles involved in conducting IO. In addition to this seminal publication, because these influence campaigns are often conducted long before the traditional beginning of active hostilities, the White House and the DoD realized that they needed better coordination. This interaction between federal agencies within the executive branch also brought about a renewed emphasis on the IO organizational structure.

Information Operations Organizations

IO by definition is normally broken down into offensive and defensive disciplines in order to better understand the relationship between different capabilities and their related activities. One can view the organizational structure of IO in the same manner, and most of the offensive capabilities of IO are retained and used by the DoD, Department of State (DoS), Central Intelligence Agency (CIA), and the White House. Although these organizations do not control all of the offensive IO capabilities of the USG, in general they tend to be responsible for the vast majority of such operations. The same, however, cannot be said of the defensive IO architecture, because these capabilities tend to be distributed out much further among the agencies, as shown in Diagram 1-9. In fact, it can truthfully be said that every organization is ultimately responsible for maximizing its own defensive posture, whether it comes in the form of information assurance, force protection, or operations security.

Therefore, the overall USG IO architecture is neither simple nor easy to understand. Relationships have evolved over a number of years, for a variety of circumstances, including political, budgetary, and perhaps even arbitrary reasons. Many organizations originally designed to conduct certain missions are currently being asked to change in this new era of interagency cooperation. For example, the Secretary of Defense under President Clinton initiated an effort to take control of the somewhat chaotic DoD IO relationships to develop in concert with other agencies a more

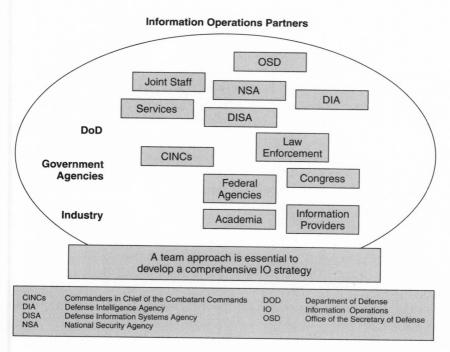

Diagram 1-9 Information Operations Partners

coherent organizational architecture.[13] Likewise, the Bush administration has also instituted a number of changes: its recent move to build a new Homeland Security Department, the stand-up of new offices in the executive branch, and a reorganization of the Combatant Commanders (CCs). All of these organizational evolutions have obviously been attempts to shift the industrial-era USG structure to a more informational age-architecture, as shown in Diagram 1-10.

Top-Level Leadership

To start at the top, the USG has always been led by civilians. The president of the United States is the senior elected official, and together with the secretary of defense forms the National Command Authorities (SecDef). Though the SecDef can initiate offensive military action, only Congress can declare war.[14] What is very interesting about IO in relation to the SecDef is that because offensive IO is often conducted before hostilities begin, the approval process for these operations often happens only at the very top of the chain of command. Therefore, it is important to keep in mind that even though many organizations may have a role in the formulation of IO strategy, policy, and tasks, the actual decision to undertake a particular offensive IO action will often come only from the SecDef in support of national-level goals. In addition to the president and secretary of defense are the vice-president and the secretary of state, who act as the statutory voting members of the National Security Council (NSC). When the NSC meets during periods of national crisis, these four people are supported by a number of other non-statutory and non-voting members.[15] The cabinet, which is composed of fifteen department heads known as secretaries, also assists the president in these executive efforts.[16] These cabinet heads are also often referred to as the Principals Committee, and they have assistant, under, and deputy secretaries who are sometimes referred to as the Deputies Committee. Together, these groups often gather in a wide variety of meetings concerning the implementation of IO at the strategic and operation levels. In addition, because IO is not limited to the DoD in its missions, other cabinet members, notably the DoS, Department of Commerce (DoC), the Department of Justice (DoJ), and the new Department of Homeland Security (DHS) also play major roles in the national IO architecture. In addition to these cabinet-level agencies, the White House has a number of different offices and agencies that are directly responsible to the president, as shown in the following discussion.[17]

Because IO is a process used to integrate operations in the information age, it will be conducted across the spectrum of conflict. Due to the continuous nature of IO, the DoD may not always be the lead agency

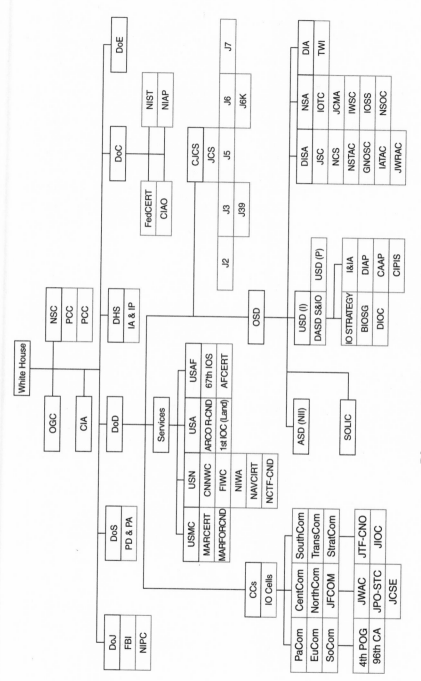

Diagram 1-10 USG IO Organizations

from the USG, and, in fact, there are many instances for which other departments such as state or commerce may be much better suited than the defense department to lead a part of the IO effort. Although the DoD doctrine for interagency operations can be found in Joint Publication (JP) 3-08, *Interagency Coordination during Joint Operations,* volumes I and II, it is much more likely that the operation may require a broader element of organizational and structural change to be truly effective in this new era. A classic example may be a nation-building mission in Central America or perhaps the development of a business infrastructure in Southeast Asia. This, fact of course, is a byproduct of the horizontal integration and cooperation that evolved between the different government agencies as the development of whole new interagency partnerships emerged.

IO and the Interagency Process

The U.S. interagency process consists of both formal and informal procedures which can be used to conduct IO missions. Established bodies such as the NSC characterize the formal interagency process, with its Principals and Deputies Committee, as well as the former Interagency Working Groups of the Clinton era, which are now called Policy Coordinating Committees in the Bush administration. These bodies attempt to coordinate from the bottom up with every effort made to resolve issues at the lowest level possible,[18] as shown in Diagram 1-11. In addition, the Clinton administration also published a number of policy documents called Presidential Decision Directives (PDDs). One of these, PDD-56, *Managing Complex Contingency Operations,* is especially important concerning the interagency process because the NSC is perhaps the only government agency that is designated to coordinate the different departments of the government. By law, each department is a separate organization and only

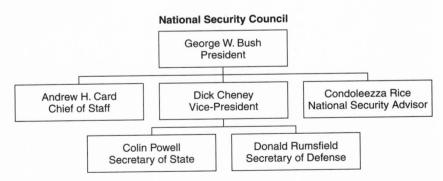

Diagram 1-11 Statutory Members of the NSC and Advisors

reports to the president. Therefore, the ability of the NSC to coordinate activities within the different departments is crucial to the overall success of any U.S. government policy. Thus, it seems apparent that in the IO world, it is the NSC that has evolved as the most important entity in the interagency process, and probably will continue to do so in the future.[19]

In addition to the formal and informal interagency process, the NSC is also involved in the promulgation of administration strategy and policy in several different methods. The first is the National Security Strategy, which was most recently published in its latest version in September 2002. This is a collaboration of many different departments and is a formal, unclassified method of addressing the security concerns of the United States throughout the world. In addition, the White House can initiate strategy and policy issues through a variety of means including presidential policy (whether the PDDs of the Clinton administration or the national security presidential directives [NSPDs] of the Bush administration), presidential determinations, findings, executive orders, presidential speeches, letters, memoranda, the State of the Union Address, press conferences, interviews, and statements by the president and other administration spokespersons. The White House can also issue policy through the use of reports to Congress and other published documents, including testimony to Congress, directives, and instructions issued by various departments and agencies. Specifically concerning IO and the interagency process, the Clinton administration used the promulgation of PDDs to form numerous groups and committees in its two terms to lead these specific interagency coordination issues:

- Peacekeeping Core Group (PDD-25)
- Counterterrorism Security Group (PDD-39 and PDD-62)
- Special Coordination Group (PDD-42)
- Executive Committees—Complex Humanitarian Emergencies (PDD-56)
- WMD Preparedness Group (PDD-62)
- Critical Infrastructure Coordinating Group (PDD-63)

Though all of the PDDs were officially cancelled by NSPD-1 in February 2001, many of the same functions and personnel have remained in position at the NSC, working on these different issues.

In addition to the NSC, there are other executive advisors involved with IO, as shown in Diagram 1-12. For example, within the Office of Management and Budget (OMB) resided the President's Council on Year 2000 (Y2K) Conversion. The council comprised more than thirty major federal executive and regulatory agencies that were responsible for coordinating

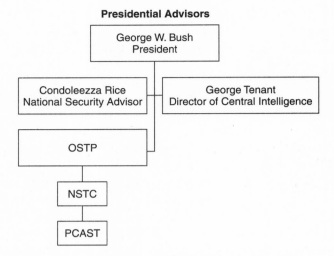

Presidential Advisors

Diagram 1-12 Other Presidential Advisors

the USG's efforts to resolve the Y2K issue.[20] Guidance for this council was amended in 1999, with the establishment of an Information Coordination Center (ICC) at the General Services Administration (GSA).[21] The ICC worked in concert with other computer emergency response teams to handle not only Y2K-related issues but also other viruses and computer network attacks (CNAs).

Likewise, another White House agency that is heavily involved with IO is the Office of Science and Technology Policy (OSTP) and its two subdirectorates, the National Science and Technology Council (NSTC) and the President's Committee of Advisors on Science and Technology Policy (PCAST).[22] These two councils act as executive advisors to the president and cabinet to coordinate the science and technology policymaking processes within the USG. Both agencies were highly successful in legislating technology-oriented issues during the Clinton administration, specifically sponsoring and recruiting for federal support of computing and communications research and development. The OSTP also sponsors the Committee on National Security, which serves as the focal point for the debate on national encryption standards.[23]

DoD—The Office of the Secretary of Defense and IO
Turning from the White House to the Office of the Secretary of Defense (OSD), the primary assistant secretaries involved in IO issues include the assistant secretary of defense for command, control, communications and intelligence (ASD/C3I) and the under secretary of defense for policy, or

DASD S&IO	
Infrastructure and Information Assurance	IO Strategy and Integration
IA Advisor to CIO	IO Policy
IA Policy	IO Security
CIP Policy	IO Program Coordination
DIAP	BIOSG/DIOC Secretariat
CAAP	IO Exercise Support
IA Planning	Future IO Concepts
DoD CIAO Support	Integration of IO and Space Control
Works closely with J-6K	Works closely with J-39

Diagram 1-13 DASD S&IO Organization

USD(P). In May 2003, ASD/C3I was reorganized into two new sub-directorates in order to better differentiate with the Title 10 and Title 50 (intelligence) policies, the assistant secretary of defense for networks and information integration (ASD/NI2) and the under secretary of defense for intelligence, or USD(I). Within both of these offices there are a number of deputy assistant secretaries (DASDs), with the most important from an IO viewpoint being the DASD for security and IO (DASD S&IO), under the USD(I). This agency is crucial because it is the one single directorate within the OSD that has both the offensive and defensive elements of IO for policy and programming. The DASD S&IO is also further divided into four different sub-directorates. Two of these, infrastructure and information assurance (I&IA) and information operations strategy and integration (IO S&I), employ most of the OSD staffers who are involved day in and day out in IO planning, policy, and strategy. To see how these interrelationships work, consider the IO mission-tasks of the OSD. Within the DASD S&IO directorate are sub-elements that coordinate across with many different agencies. One good example is the Critical Infrastructure Protection (CIP) policy branch, whose personnel coordinate daily with the CIAO and the NIPC, both of which are explained in detail later in this chapter. Elements of the DASD S&IO organization are shown in Diagram 1-13.

Thus we can see that with this most recent reorganization, the USD(I) retained most of the IO branches that were formerly a part of the ASD/C3I. This is important, because now there is a direct correlation between intelligence and IO, with both the offensive and defensive portions of policy collocated. The other directorate of OSD, the USD(P), is also now under USD(I) and is similarly heavily involved with IO policy and doctrine. Most of its authority was originally derived from PDD-29

Security Policy Coordination.[24] This revision of the security policy process was needed to help give the United States greater security, given a wider diversity of threats in the post-Cold War era. The USD(P) also has a number of sub-directorates that involve IO policy issues, including who coordinates processes such as PDD-68, which is mentioned later in the book. However, with both the DASD S&IO and USD(P) having major advocacy of IO policy within the OSD, there was bound to be confusion at times between these two organizations. Therefore, in a memorandum of understanding between USD(P) and the ASD/C3I during 1999, it was agreed that the USD(P) would have the policy lead on development and oversight of offensive IO, psychological operations (PSYOPS) and international public information (IPI).[25] It was also agreed that the assistant secretary of defense for special operations and low-intensity conflict (ASD-SOLIC) would retain IO tactics, techniques, and procedures that are unique to special operations forces, and in turn, the ASD/C3I would have the lead for policy development and oversight of information assurance (IA) as explained later in Chapter 3. This architecture has been basically retained with the setup of the ASD/NI2 in 2003.

DoD—Combat Support Agencies

The Secretary of Defense is also supported by combat support agencies, of which there are nine. The three main staffs involved in IO are the National Security Agency (NSA), the Defense Information Systems Agency (DISA), and the Defense Intelligence Agency (DIA). Each of these organizations, in addition to conducting their typical support for traditional military operations, has also formed new units to support the unique needs of the IO structure. The first two of these organizations will be discussed in this section, and the DIA will be covered in the section discussing the intelligence community.

The NSA's IO Architecture

The NSA is the primary U.S. intelligence agency officially designated to conduct signals intelligence (SigInt).[26] Its charter takes many forms, most of which are highly classified, but clearly the NSA is very interested in the increase in computer hacker activity in the United States. Because it needs to monitor computer network defense (CND) issues as much as, if not more than, other agencies, it is not surprising that it formed its own computer emergency response team (CERT). Titled the National Security Operations Center/Information Protect Cell (NSOC/IPC), this organization is a separate cell that stands a 24/7 watch to monitor SigInt and information security incidents in order to protect NSA's networks from attack.

Tied to this NSA CERT is another organization, the National Security Incident Response Center (NSIRC), which operates as an analysis center. This entity shares incidents and threat vulnerability information with all USG departments and agencies as well as with its contractors involved with the national security strategy.[27] The NSIRC develops the National Information Systems Weekly Incident Summary, which contains national-level, operationally fused data that correlates computer incidents and events that might formerly have been viewed in isolation by a uniformed service, CERT, or agency. NSA also participates in the National Security Telecommunications and Information Systems Security Council (NSTISSC), which was established as a senior-level policy coordinating committee to consider technical matters and policies. Its members include personnel from ten different government departments and its primary mission is to protect national security systems.[28] In addition, the council also supports IA, information security, and CND training, with a scope limited to national security information and systems. A final command under the NSA umbrella is the Joint COMSEC Monitoring Activity (JCMA). With detachments deployable around the globe, the JCMA performs information security monitoring, analysis support, communications security, and cryptographic monitoring.[29]

The DISA's IO Architecture

The DISA, like the NSA, is also located in the greater Washington, D.C. metropolitan area. Chartered to maintain and protect the majority of the DoD's computer networks, DISA is very concerned with the detrimental effects of CNA efforts against the United States. It has had a CERT capability for a long time, namely its Global Network Operations Security Center (GNOSC), which monitors the operational and security posture of the defense information infrastructure (DII). Although DISA does have a direct tie to the uniformed services for CND efforts, it does not have command authority to direct a military service to change network configurations or settings. Therefore, in 1998, after a series of well-publicized attacks on USG computer networks, the OSD directed that DISA set up the Joint Task Force-Computer Network Defense (JTF-CND).[30] This command directly communicates with the other government agencies including the DoJ, the federal CERT, and NSA's CERT and the various Service CERTs. Transferred to operational control of U.S. Space Command (SpaceCom) in 1999, the JTF-CND is in effect the senior military CERT and the DoD response cell for CND issues, including recommending changes to the information condition (InfoCon) status when the situation dictates. In April 2001, this organization was renamed the Joint Task

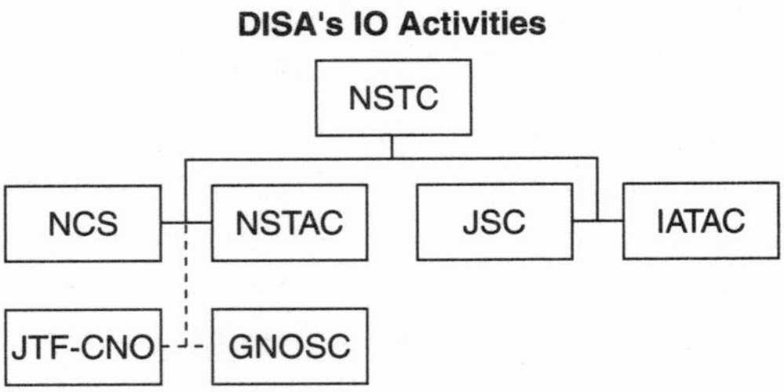

DISA's IO Activities

Diagram 1-14   DISA's IO Activities

Force-Computer Network Operations (JTF-CNO) to reflect its growth and mission, and continued to operate under the IO portion of the Space-Com mission, although still physically collocated with DISA, and still receiving administrative support from them as well. On January 10, 2003, U.S. Northern Command (NorthCom) was set up to coordinate military homeland security efforts, and thereby U.S. Strategic Command (Strat-Com) also modified its mission as well. The former SpaceCom was absorbed into the two other CCs, with the IO tasking going to StratCom in Omaha, Nebraska. Therefore, the chain of command for the JTF-CNO was also shifted to as well, this time from SpaceCom to StratCom.

A number of other organizations within the larger DISA umbrella have existed for years, but are now being adapted to perform IO missions. These include the National Communications Systems (NCS), the National Security Telecommunications Advisory Committee (NSTAC), and the Joint Spectrum Center (JSC), as shown in Diagram 1-14. The NCS was created to ensure governmental communications after problems occurred during the Cuban Missile Crisis. Comprised of twenty-three federal agencies and the telecommunications industry, the NCS maintains a coordinating center to resolve failures of the public switching network,[31] as shown in Diagram 1-15. In addition, the NCS also operates an High Frequency radio system, independent of the public switched network, to provide connectivity to the Federal Communications Commission, regional Bell, GTE, Sprint, and switch manufacturers.

The NSTAC was created in the aftermath of the American Telephone & Telegraph divestiture and serves as a forum for addressing the risks to U.S. national security posed by potential threats to national telecommunications and information industries.[32] It represents a joint government/ industry partnership, the likes of which have not been seen since World War II. Comprised of thirty chief executive officers from the telecom-

National Communication System Members	
Department of State	Joint Chiefs of Staff
Department of the Treasury	General Services Administration
Department of Defense	United States Information Agency
Department of Justice	National Aeronautics and Space Administration
Department of the Interior	Federal Emergency Management Agency
Department of Agriculture	Federal Communications Commission
Department of Commerce	Nuclear Regulatory Commission
Department Health and Human Services	National Telecommunications Information Administration
Department of Transportation	Federal Reserve System
Department of Energy	National Security Agency
Department of Veterans Affairs	United States Postal Service
Central Intelligence Agency	

Diagram 1-15 National Communication Systems Members

munications, information technology, aerospace, and banking industries, NSTAC makes recommendations to the president on issues critical to protecting the U.S. communication infrastructure; thus, its role grew in importance due to Y2K issues in 1998–1999. This committee also boasts a fifteen-year string of successes, including the establishment of the National Coordination Center for Telecommunications, the Network Security Information Exchange, and the Government Emergency Telecommunications Service. In addition to these public-private partnerships, DISA also has purely military units that have major roles in coordinating IO products. One example of this is the Joint Spectrum Center in Annapolis, Maryland, which is an outgrowth of a need to coordinate frequency spectrum management. It also assists in the development of the joint restricted-frequency list, and the resolution of operational interference and jamming requests.[33]

A final command under the DISA community that is worth mentioning is the Information Assurance Technology Analysis Center (IATAC). This center is responsible for a number of functions and tasks, but the one that is probably most useful to the IO operator or planner is the education role. IATAC has done a magnificent job over the last few years in making and distributing a wide variety of IO and especially IA teaching tools. Some of these are soft copy, others are distributed as CD-ROMs, but nonetheless, they are invaluable to helping IO professionals complete their missions.[34] See Diagram 1-16 for a list of IATAC products.

IATAC Products
Defense in Depth
Vulnerability Analysis Tools Report (2nd ed.)
Data Mining
Data Embedding for IA
Instrusion Detection Tools Report (2nd ed.)
Computer Forensics—Tools and Methodology
Malicious Code Detection
Modeling and Simulation Technical Report
Biometrics: Fingerprint Identification Systems
IA Metrics
Firewalls
Visualization

Diagram 1-16   IATAC Product List

DoD—The Joint Staff and IO

From the uniformed military perspective, the secretary of defense is supported by the Joint Chiefs of Staff (JCS), which is comprised of a senior military officer from each branch of service plus a chairman and vice-chairman. These officers, in particular the chairman of the Joint Chiefs of Staff (CJCS), act as the principal military advisors to the SecDef. Although the JCS does not "own" or actually command troops in combat, nonetheless its advice, more often than not, has a great effect on the armed forces. The Joint Staff supports the JCS, and it is organized along typical U.S. military doctrinal terms, with J-3 being operations and J-39 being the deputy director for information operations (DDIO). The J-39/DDIO is responsible for IO doctrine and has authored the baseline DoD document, JP 3-13, *Information Operations*.[35] If there were only one office as the central point for IO in the Pentagon, then J-39 would be it, for it is the primary JCS organization that interacts with OSD staffers (specifically DASD S&IO) and it is also the liaison to the JCS for each CC's IO cell.

In a coordination role, you will also see the J-39/DDIO staff working with a number of different DoD staffs such as ASD/NI2 and USD(P), as well as other USG and interagency commands, to ensure continuity of IO plans and doctrine. But because so much of IO is really nothing more than detailed integrated planning, the JCS is normally not tactically involved with each and every CC IO cell plan. Instead, JCS will attempt to stay focused on the broader issues, such as those involving IO policy, strategy, and doctrine. Thus there is no such entity as a "CINC" IO. Although sev-

eral unified command plan (UCP) proposals have illuminated this deficiency, to date, no new command or sub-unified agency has been formed.
Instead, all CCs have emphasized their particular specialties and capabilities associated with IO to develop inter-related working cells. A good
example of this is Special Operations Command (SOCom), which is the
combatant command of the PSYOPS and civil affairs forces for the U.S.
Army. However there still is a need for more expertise, especially with the
emergence of CND and CNA as warfare areas, as few personnel on CC
staffs typically have identifiable skills in these types of operations.

In the offensive-defensive IO terms mentioned earlier, it would be
preferred to have staffs represent both sides of the warfare spectrum for
IO, and that is what the DoD has done. Thus, though J-39/DDIO is a full-
spectrum staff for IO, it is also primarily an offensive-oriented organization.
On the defensive side for the JCS are the J-6 organization, or the command,
control, and communications (C3) department. Specifically for IA or CND,
J-6K has been designated as the responsible staff to deal with these asymmetric threats.[36] To do this, it has maintained close liaison with the DASD
S&IO as well as the service CERTs, JTF-CNO, and CC's J-6s.

The CCs

The real locus of operational-level planning for IO is usually with the military CC's and their IO cells. It is the CCs who are often engaged in IO on
a day-to-day basis. IO planners on the CC's staff use the NSS and national
military strategy (NMS) as their guide to outline in broad terms the CC's
operations plans and theater engagement plans. These CC IO cells are also
involved in the day-to-day operations that are not necessarily directly
combat related. For example, the earlier 1999 NSS chartered the CCs to
plan to conduct a variety of operations including non-combatant evacuation operations, special forces assistance to nations, humanitarian and disaster relief, and so on. But probably most important is the daily overseas
presence mission that encompasses a host of operations that takes the U.S.
military into areas far beyond its traditional bases. In addition, the task of
supporting other national objectives also brings the United States into
operations all over the world. It is therefore crucial that these CC planners
integrate these operations not only with the respective services, but with
other executive department counterparts as well.

In the U.S. military, CCs are also the actual commanders that "own"
military forces. There are nine CCs, of which five are regional:

- Central Command (CentCom), Tampa, Florida
- European Command (EuCom), Stuttgart, Germany

- Pacific Command (PacCom), Honolulu, Hawaii
- Southern Command (SouthCom), Miami, Florida
- Northern Command (NorthCom), Colorado Springs, Colorado

The other four are functional and conduct their missions across the globe:

- Joint Forces Command (JFCom), Norfolk, Virginia
- Special Operations Command (SOCom), Tampa, Florida
- Strategic Command (StratCom), Omaha, Nebraska
- Transportation Command (TransCom), St. Louis, Missouri

This structure is relatively recent, with the latest changes coming in the summer of 2002 with the announcement of the set-up of Northern Command or NorthCom. Developed in response to the attacks of September 11th, this CC will coordinate homeland security issues, and will work closely with the Department of Homeland Security (DHS) as well as North America Air Defense (NORAD). Based in Colorado Springs, NorthCom will take over many of the tasks previously assigned to SpaceCom; the major IO role that SpaceCom once had has now been shifted to StratCom in Omaha, Nebraska. Yet this one CC does not do all IO for the nine CCs. Instead, within each of these CCs is still an IO cell, which is part of the staff. Typically, these CC IO cells are very small in manpower, but they can expand during actual contingency operations or planning. Each cell is responsible for the detailed IO planning done for its particular CC; however, it also normally coordinates and works in conjunction with the J-39 division of the Joint Chiefs of Staff, which has overall responsibility for IO, as discussed earlier.

Outside the strategy and policy arena, a number of DoD organizations have been formed in the last few years or have evolved from older legacy commands agencies. In addition, some of these commands were classified at levels higher than Top Secret and emerged only with the recent downgrading of the classification of certain IO terms. Originally a number of these agencies worked directly for the Joint Chiefs of Staff or JFCom. However, that changed with the Unified Command Plan 1999 (UCP '99), which gave SpaceCom the lead in CND, effective October 1, 1999, and CNA, effective October 1, 2000.[37] SpaceCom acted as the supporting CC to the other commands for these missions and also coordinated with the Joint Chiefs of Staff on these issues. With UCP '99, JTF-CND and the JIOC were reassigned to SpaceCom.[38] As a counterpart to the CERT set up at the FBI headquarters (NIPC), as mentioned earlier,

JTF-CND was originally designed to be a small staff (twenty-four person-nel) who would stand watch and analyze the implications of a military failure on the network or system. Therefore, not only did the original con-tingent include computer experts and military lawyers, but there were also operators, including fighter pilots, among the staff.

As previously mentioned, on October 1, 1999, JTF-CND was assigned to SpaceCom as part of the transfer of the CND mission to that CC. Staffing has been somewhat increased with a number of allied officers detailed to the command. In addition, there is also discussion of forming an international JTF-CND with Australia, New Zealand, Canada, and the United Kingdom as primary members. Most of the impetus for this shar-ing actually came from the United States. When the "ILOVEYOU" virus attacked computer systems in 2000, it was the Australia and New Zealand commands that were the first to know, but releaseability issues hindered the notification of other allies. To date, JTF-CND has had a pretty good track record. The use of Information Conditions (INFOCONS) by the DoD, an original function of the command, has been well received—in many cases, JTF-CND gets more respect as a CERT than the NIPC. This is due to a variety of reasons, but most important may be the willingness to handle all agencies fairly and without a political agenda.

During the winter of 2000–2001, additional changes occurred in the organizational structure of SpaceCom with respect to IO. It was during this period that the deputy director of operations for computer network attack (DDO-CNA) office was formed. This small staff of seven individu-als was developed to support SpaceCom in its efforts to advocate CNA within the Pentagon, as well as to facilitate the approval process. As men-tioned earlier, in February 2001, this group merged with the JTF-CND to form the Joint Task Force—Computer Network Operations (JTF-CNO) to better meet the needs of SpaceCom to conduct IO. Other changes occurred as well. On April 2, 2001, JTF-CND changed its name to JTF-CNO to better reflect its missions and operations. Once again, these changes emphasize the continual evolution of the organizational architec-ture of IO, and finally, as mentioned earlier, with the set-up of NorthCom, all the IO activities mentioned within the last few paragraphs have been transferred to StratCom.

Additional DoD IO Elements
In addition to J-39/DDIO, OSD, and CC IO cells, a number of other "play-ers" or agencies also have a piece of the IO pie. First, the intelligence com-munity (IC) is made up of many diverse agencies, such as the CIA, DIA, and NSA, which have a long history of involvement in capabilities normally

associated with IO, such as operations security (OpSec) and military deception. Originally many of these organizations were formed to conduct a certain mission or operation and not intended to interact as they are currently being asked to do. They were stovepipe agencies or legacy commands that reported vertically up and down the chain of command. Now, because of IO and the urgent need to have interagency cooperation, especially in the wake of September 11, 2001, it has become much more common for all of these unique organizations to work together. In fact, most of the CCs have a number of permanent intelligence representatives assigned to act as agency liaisons for IO missions. The intelligence community is also heavily involved in supporting operational requirements for IO with permanent seats on a number of interagency groups and committees. Of particular importance early on in the IO developmental period of 1998–2000 were the Bilateral IO Steering and Working Groups (BIOSG/BIOWG) that helped to define IO policy and deconflict IO issues between the DoD and other agencies. Typical members of the BIOWG were at the one-star level, and these members defined the issues and laid the groundwork for the BIOSG, which actually made the decisions and wrote policy at the three-star level to include representatives from the OSD, the Joint Staff, and the IC.

Cabinet IO Interests
Other departments besides the DoD also have vested interests in IO. Because of the global and over-arching role of IO, agencies such as the DoS, DoC, and the DHS DoJ have begun to play roles that are much more important in IO, especially in the defensive arena. Much of IA is defined in business or legal rather than military terms; therefore, it is only natural that these organizations have begun to carve out their niches in the IO structural architecture.

Department of State IO Concerns
In the foreign policy arena, the State Department (DoS) is the major organization that conducts diplomacy for the United States around the world. With the need to present a coherent public affairs and information front to the international media, the DoS was reorganized in 1999 to bring the formerly independent United States Information Agency (USIA) into its larger umbrella organization. Renamed as the under secretary of state for public diplomacy and public affairs, this new directorate now coordinates both international public information (IPI) and PA areas within the DoS.[39] Although both of these areas are discussed later in Chapter 4, it is important to note how crucial they both are to the success of an IO campaign. This was evidenced by the publication of PDD-68, *International Public*

Information, during the middle of the Kosovo campaign in 1999.[40] To win the hearts and minds of an enemy, and to achieve one's operational and strategic goals, you must be prepared to influence foreign audiences with a coherent message.

Traditional DoS Structure

The interaction between the CC and foreign nations relies heavily on the ambassador and the country team. State department representatives are essential to successful operations and have broad powers. The key members include:

- DoS regional secretary
- Ambassador
- Political advisor
- Country team
- Resident military representative

The ambassador and the country team have several documents and policies that they use to plan their operations within their area of interest. These policies and programs are important to the interagency process because they must be taken into consideration in any CC's theater engagement plan (TEP) or operations plan. These policies include the DoS regional program plan (RPP), which defines regional and country objectives and strategy. The DoS RPP is prepared by the regional assistant secretary and is a product of the interagency process, which reflects the international affairs strategic plan. At the embassy, the mission program plan (MPP) is prepared by the country team and is the ambassador's country engagement plan. Of special notice to military planners, the CC's TEP should consider all MPP's of interest in their area of responsibility (AOR). These documents are readily available to the CC's planners and can be found in the DoS's Congressional Presentation for Foreign Operations. The MPPs are important because they contain measures of effectiveness, objectives, and priorities for the State Department in support of the NSS.

DoC IO Architecture

The State Department is not the only cabinet-level agency that is changing under the influence of IO. The DoC has also played a major role in IO over the last five years. One of the reasons for a cabinet agency that is primarily concerned with business and finance to be involved in this new warfare area is because the DoC is heavily involved in the second of the two new capabilities, namely computer network operations (CNO). Although the

DoS, as mentioned earlier, has a huge role in perception management, DoC also has an equally important mission concerning CNO, particularly with regard to information assurance (IA).

The Commerce Department is also the host agency for the CIAO, the sub-directorate agency that was established as a direct result of the proclamation of PDD-63, *Critical Infrastructure Protection (CIP)*.[41] The CIAO is officially tasked to coordinate CIP within the USG, and it evolved from the Presidential Commission on Critical Infrastructure Protection (PCCIP). This group was comprised of government officials, commercial businessmen, military and civil service personnel, and academics. These executives met over an eighteen-month period in the 1996–1997 timeframe and produced a document called *Critical Foundations,* which linked CIP to national security and identified eight critical industries, as follows:[42]

- Telecommunications
- Electrical power systems
- Gas/oil storage
- Banking/finance
- Transportation
- Water supply systems
- Emergency services
- Continuity of government

These industries are essential to the economic and security infrastructures of the United States, and the publication of *Critical Foundations* led directly to the formulation of PDD-63 in May of 1998.[43] Tied into a larger Clinton administration effort of interagency IO efforts, PDD-63 had a sister directive, PDD-62, *Counter-Terrorism*, which was published at the same time, and both documents came under the authority of the NSC (discussed in further detail in Chapter 3). The key to the success of PDD-63 was the fact that it took the eight industries identified as crucial to the security of the nation and then tied them to a cabinet department as well as to a comparable private industry association. Together, then, the USG and private industry have since produced a CIP plan to work together to protect these resources from attack.

In addition to the CIAO, the DoC also hosts the National Institute of Standards and Technology (NIST) and the National Telecommunications and Information Administration (NTIA). The mission of the NIST is to promote economic growth around the world, and it does this by working with private companies to develop and apply technology, measurements, and standards.[44] Specific tasks include the following:

- Assist industry to develop technology to improve product quality.
- Modernize the manufacturing process.
- Ensure product reliability.
- Facilitate rapid commercialization of products based on new scientific discoveries.
- Develop information system security guidelines, procedures, and technological solutions to help federal agencies implement OMB policy.[45]

The current areas of interest for the NIST include electronic commerce, public key encryption, common criteria for information technology, advanced authentication, and the Federal Computer Incident Response Cell (FEDCIRC). NIST also hosts the Federal Agency Computer Security Program Managers' Forum, which advocates information exchange on information technology issues.[46] The forum cannot command or regulate changes, but instead is mainly used as an information-sharing group. In addition, NIST also collaborates with the NSA in the National Information Assurance Partnership (NIAP). This organization was designed to combine the extensive computer security experience of both the DoC and the NSA.[47] NIST is also the host for the Information Infrastructure Task Force (IITF), which works with the private sector and government agencies under the cognizance of the OSTP. The National Technological Information Association is also a NIST organization and is the principal voice of the executive branch on domestic and international communications and information technology issues.[48] Specifically, NTIA was involved in the Telecommunications Act of 1996 which eliminated many barriers to ownership and operation in the telecommunications and broadcast industry, making private ownership far easier than before. The relaxing of requirements has made it harder to secure and control the National Information Infrastructure (NII), but one has to ask wonder if the internet should, or could, be controlled. Finally, the DoC also hosted the USG Y2K Task Force, which was an offshoot of the PCCIP process and the CIAO.

DoJ IO Architecture

The other organization recommended in *Critical Foundations* was an information-warning center. Although similar in concept to the CERTs (which were already in existence), it was envisioned that this new legal center would use Federal Bureau of Investigation (FBI) expertise to prosecute cybercrimes at a national level. In 1998, the FBI formed the NIPC, which was charged to maintain liaison with law enforcement personnel throughout the nation, as well as with all fifty-six FBI field offices.[49] NIPC is also tied into the CND arena with contacts at the JTF-CNO and NSA. Together, this allows the executive branch to use its legal authority under the FBI and DoJ to prosecute cyberterrorism within the United States.

Department of Homeland Security

In the summer of 2002, President Bush reiterated his intentions to build a new cabinet agency, the Department of National Homeland Security, for which former Governor of Pennsylvania Tom Ridge was selected to lead the development of, starting as early as September 20, 2001. Two weeks later, on October 8, 2001, the President signed Executive Order 13228, which established the Office of Homeland Security and the Homeland Security Council. However, it was quickly recognized that without budget line authority, this organization would not be powerful enough. Therefore on October 11, 2001, in a bipartisan effort, Senators Joseph Lieberman (D-CT) and Arlan Specter (R-PA), sponsored a bill to create a new cabinet-level agency that would have fiscal responsibility over a vast array of current government organizations, that the then-current structure developed under the executive order, did not. Under the Senate plan, the Department of National Homeland Security was to be organized into three functional parts, focused on prevention, protection, and preparation. Altogether, eight organizations or offices were supposed to be transferred to this new directorate:

- Coast Guard (USCG)
- Customs Service (USCS)
- Border Patrol
- Commerce's Critical Infrastructure Assurance Office (CIAO)
- Commerce's Information Infrastructure Protection Institute
- FBI's National Infrastructure Protection Center (NIPC)
- FBI's National Domestic Preparedness Office
- Federal Emergency Management Agency (FEMA)

What is very interesting from all these proposals is that much of the current legislation mirrors almost exactly the reforms suggested by President Clinton's Blue Ribbon Commission on National Security in the twenty-first century. Led by former Senators Gary Hart (D-CO) and Warren Rudman

(R-NH), this report was released in three parts earlier in 2001, and suggested the creation of a new cabinet-level agency. Nothing much was done on the bipartisan proposal by the White House until the summer of 2002, when President Bush formally announced the creation of a new cabinet-level agency, the Department of Homeland Security. Once again, sweeping in nature, this new organization now includes more than 170,000 government employees, whose major missions consist of the following tasks:

- Border and transportation security
- Emergency preparedness and response
- Chemical, biological, radiological, and nuclear countermeasures
- Information analysis and infrastructure protection

Because this department is so new, it will take time to determine if this new organization will be effective in its efforts. However, that said, there is still much to be done now with the current war on terrorism.

Interagency IO Organizations

The fact that IO requires significant horizontal integration is a very important fundamental concept with these different cabinet agencies. Numerous interagency groups and councils have been formed in the last five years to facilitate the much-needed integration of IO activities in the United States. Some have been mentioned previously but the following are examples of those that were created during the Clinton administration:

- Bilateral Information Operations Steering Group (BIOSG)
- Bilateral Information Operations Working Group (BIOWG)
- Critical Infrastructure Protection Working Group (CIPWG)
- Defense Information Assurance Program Steering Group (DIAPSG)
- National Information Assurance Partnership (NIAP)
- National Science and Technology Council (NSTC)
- National Security Telecommunications Advisory Committee (NSTAC)
- National Security Telecommunications and Information Systems Security Council (NSTISSC)
- Office of Science and Technology Policy (OSTP)
- President's Committee of Advisors on Science and Technology Policy (PCAST)
- International Public Information Interagency Working Group (IPIIWG)

Some of these organizations were mentioned earlier in the DoD section, but there are also other IO-related groups or councils in the OSD, including the Defense Information Operations Council (DIOC) and the DIAPSG. These organizations were both three-star working groups that tried to coordinate and deconflict IO issues within the DoD. A final interagency working group that is a holdover from the Clinton administration is the Forum of Incident Response and Security Teams (FIRST). Hosted by the Department of Energy (DoE), this forum has a long history of working with various CERTs to combat computer viruses and attacks.[50]

Interagency coordination involves working with more than just organizations originating from the USG. Academia, private industry, and coalition governments are also crucial for the development of true interagency operations. This can be seen in the Operation Noble Anvil campaign in Kosovo during 1999. It was here that the utility of working not only in the joint world but also in the combined world with other nations and organizations demonstrated how crucial horizontal interaction can really be. Yet there are still even more organizations that play roles, including private or commercial agencies which may be involved in one form or the other in interagency operations. These include non-governmental organizations (NGOs) and private voluntary organizations (PVOs), a few of which follow:[51]

- Concern Worldwide Limited
- International Organization for Migration (IOM)
- Medecins Sans Frontieres (MSF; Doctors without Borders)
- OXFAM
- Save the Children

PVOs are non-profit humanitarian assistance organizations involved in development and relief activities. In the last few years, this term has disappeared from academic literature and most of these organizations are routinely called NGOs as well.

Probably more than anywhere else, in an IO mission it is crucial that the CCs understand and appreciate the importance of these NGOs. These organizations are crucial to the success of that mission when conducting IO during peacetime or in military operations other than war (MOOTW). Often the NGOs can operate where uniformed military personnel cannot, and they can often gain the trust of the locals much better than any USG agency. In addition, NGOs may have capabilities including communications, transportation, public affairs, and medical facilities that rival or surpass those available to a CC in a particular area. It is therefore in the CC's best interest to be actively engaged and to work closely with the NGOs in their area of operations. As a CC or U.S. military planner, one cannot command or direct NGOs to conduct missions. Instead, what normally works best is to facilitate these agencies and to work with them in order to conduct one's operation. Again, horizontal integration is the key to success for IO.

Summary

In conclusion, there are clearly a large number of "players" in the IO arena, and trying to understand how they all relate can be quite complicated. Much of this organization is relatively new and, in fact, has changed considerably in the last few years. However, throughout this discussion of national IO organizations, the one overriding theme to remember is that for IO to succeed there must be cooperation between all parties involved. This means horizontal as well as vertical integration and cooperation, and includes not only USG agencies and departments, but also non-governmental units and private industry as well. Because so much of IO now crosses old departmental boundary areas, it is important to realize the power of information and that IO encompasses much more than the traditional DoD missions and policies. Therefore, if the United States is to succeed, it must coordinate its actions with all of the players involved—only through cross-departmental communication flow by all organizations will IO become the true force multiplier that it has the potential to be.

CHAPTER 2

Intelligence Support

Foundations for Conducting IO

"Know the adversary and know yourself, and in a hundred battles you will never be in peril. When you are ignorant of the enemy but know yourself, your chances of winning and losing are equal. If ignorant of both your enemy and yourself, you are certain in every battle to be in peril."[1]

Sun Tzu

Intelligence is the bedrock of IO. It is both foundational and essential to all military operations and its importance to influence campaigns is crucial. Skeptics may simply turn to the executive summary of JP 3-13 for proof, where the following statements appear within the first four paragraphs of the text:

"Intelligence and communications support are critical to conducting offensive and defensive information operations."

"Intelligence support is critical to the planning and execution and assessment of IO."

"Intelligence preparation of the battlespace is vital to successful IO."[2]

Intelligence is also a key element of information superiority. As shown in Diagram 1-4, one of the components of information superiority is relevant information. There is some discussion within the intelligence community on the distinctions between "information" and "intelligence." JP 2-0, *Doctrine for Intelligence Support to Joint Operations,* defines intelligence as "information and knowledge about an adversary obtained through observation, investigation, analysis or understanding," as well as "the product resulting from the collection, processing, integration, analysis, evaluation and interpretation of available information concerning foreign countries or areas."[3] Thus the key factors in determining the intelligence value of information are its "relevance" to the current military operation and its "applicability" to answering a commander's critical information requirements.[4]

The Application of IO

This section explains how intelligence supports information operations. It describes the intelligence cycle and how intelligence products get to the consumer, how the intelligence community is structured to support IO, intelligence support for IO planning, the joint intelligence preparation of the battlespace process, and finally some of the unique challenges that JV 2010/2020 bring to the intelligence community.

Current JCS policy guidance on IO is set forth in CJCSI 3210.01, which states: "Intelligence requirements in support of IO will be articulated with sufficient specificity and timeliness to the appropriate intelligence production center or other intelligence organizations to meet the IO demand."[5] What this means is that one must know exactly what one wants and one must know how to get it. An oft-heard cry in the IO field is, "I don't know enough to ask the right questions!" Like ships in the night, the IO and intelligence players are often just missing each other because they lack the ability to articulate exactly what it is they want to do or can do to support the commander's mission. The way this often plays out is that intelligence producers "push" a lot of intelligence products to the consumers in the IO community who "pull" down what they want and discard the rest.[6] The formation of IO cells and the constant interaction of the J-2 and J-3 IO players on a CC staff have done much to improve communication at the operational level, but there are still weak links at the strategic level between the intelligence and operational communities, as well as the diplomatic (DoS) and military (DoD) intelligence communities. The movement of the DASD S&IO to the new ASD/NI2 directorate in 2003 was an attempt to correct this deficiency in the OSD staff.

Joint Publication 2.0, *Joint Doctrine for Intelligence Support to Operations* (March 2000), defines the central principal of intelligence as "knowledge of the adversary." Clausewitz stated it this way: "By 'intelligence' we mean every sort of information about the enemy and his country—the basis, in short, of our own plans and operations."[7] Sun Tzu has one of the most famous quotes (see chapter opening quotation) on what it means to "know the adversary," and, as noted earlier, is considered by many to be the first "information warrior." Therefore, it is the fate of the intelligence professionals to know and understand the adversary's capabilities, limitations, and intent. This obviously is not an easy task. At some point, practically every intelligence professional must give his or her "best shot" and inform the commander of the adversary's anticipated course of action. General Colin Powell, former chairman of the Joint Chiefs of Staff, said it best when he stated, "Tell what you know . . . tell what you don't know . . . tell me what you think . . . always distinguish which is which."[8] As discussed further in the following chapters, identifying the adversary in the Information Age is problematic. Intelligence officers trained to produce doctrinal templates of a Soviet Motorized Rifle Regiment in the attack find it much more difficult to "template" a hacker attempting a computer network attack. In addition, with the target of IO being the adversary decision-maker, the need for much more detailed intelligence profiling and human factors analysis grows exponentially.

As Sun Tzu's dictum reminds us, knowing the adversary is only part of the equation. The second part is knowing oneself, or for whom one works. At times, it may be easier to collect intelligence on the adversary than it is to get the commander's attention and have him or her articulate those desires as concrete requirements.[9] Each commander comes into the position of authority with biases and preconceived notions of what intelligence can and cannot do. Added to that, the skepticism that some senior officers have toward IO makes it extremely difficult to gain a clear indication of the commander's intent and expectations when it comes to producing intelligence that will support IO.

A good intelligence officer must be able to convince the commander that intelligence is more than simply a combat and power multiplier. Intelligence will help the commander focus combat power and resources and will help provide force protection. Intelligence also helps a commander identify and determine objectives and plan the conduct of operations.[10] The J-2 intelligence officer on a joint or combined staff plays a key role in the deliberate and crisis action-planning process by producing the intelligence analysis needed to "wargame" courses of action and recommend operations to the commander. Having the J-2 officer involved in the planning process early on helps to focus the available intelligence resources on the critical information requirements for a particular operation and identifies intelligence gaps that will need to be reported to supporting agencies.

The Intelligence Cycle

The process by which information becomes intelligence and responds to the commander's requirements is the *intelligence cycle*. Within the joint community, "the intelligence cycle provides the basis for common intelligence terminology, tactics, techniques, and procedures."[11] The intelligence cycle is a conceptual model composed of six phases: planning and directing; collection; processing and exploitation; analysis and production; dissemination and integration; and evaluation of feedback.

The first phase, planning and directing, involves the identification of intelligence requirements and available resources. During this step, intelligence annexes are prepared, personnel support are identified, and coordination is effected between staff sections within the command and other agencies outside the command. The collection manager begins to formulate a collection plan that allows for the coordination of all intelligence assets, both organic ones and those of higher or adjacent units. Intelligence requirements that cannot be satisfied by organic resources must be communicated as requests for information (RFI) to other units or agencies.

The second phase is collection. During this phase, requirements have already been paired up with capabilities, and actual collection of information is taking place. The intelligence community has a large number of disciplines to tap into when trying to collect information. The following are the seven major and subordinate intelligence disciplines:

- ImInt: imagery intelligence
- SigInt: signals intelligence
- HumInt: human intelligence
- MasInt: measurement and signature intelligence
- OsInt: open-source intelligence
- TechInt: technical intelligence
- CI: counterintelligence[12]

A good collection plan will ensure that there is redundancy built into the process so that appropriate cross targeting of disciplines can occur. In other words, if ImInt identifies what appears to be a new command and control facility, then SigInt or HumInt assets could be refocused on that location to confirm or deny the existence of such a structure. The collection plan must also be synchronized with the operation plan so that the intelligence indicators will be there in time to allow the operational commander to make the appropriate adjustments to the plan. For example, identifying a named area of interest, a bridge for example, and placing collection assets on that point may help to confirm or deny an adversary's intentions to move its forces through that junction. Those indicators need to be provided in a manner timely enough to allow the commander to react.

The third phase in the cycle is processing and exploitation, wherein raw information is converted into a product that can be used by the intelligence analyst. Processing depends on command, control, communications, and computers (C4), as this is the transmittal means (links and nodes) that gets the information to the analyst. Collection is only as good as the means to get the information where it needs to go and in a usable format. For example, captured documents or open-source information may need to be translated first before going to an analyst who will evaluate that new information against present holdings.

The fourth phase is analysis and production, wherein processed information or "raw intelligence" is turned into usable intelligence products. At the joint commands, this step occurs in the Joint Intelligence Center.[13] Here, analysts having regional or functional production responsibilities (like Korean Order of Battle) fuse all-source intelligence information into an intelligence product that aims at satisfying a commander's priority intelligence requirements (PIR).

Intelligence products can take many forms. They can be bound into hard-copy reports, displayed on maps or as images, and/or reproduced electronically. Generally they fall within the following six categories:

- Current intelligence
- Indications and warnings
- General military
- Target intelligence
- Scientific and technical
- Counterintelligence[14]

The categories correspond to the purpose for which the intelligence product was produced. There is often overlap because the data can frequently be used by a number of different consumers with different specific needs.

The fifth phase is dissemination and integration, in which the final product is transmitted to the customer. Former DIA Director Rear Admiral (RADM) Wilson describes this as "the hardest part of the intelligence cycle to get right," because intelligence is of little value if it does not get to the intended recipient "at the right time, in the right format, at the right amounts, in the right place."[15] Means of transmission can include verbal reports, written documents, video teleconferencing, electronic databases, graphic products, and so on. Dissemination can occur through either a "push" system (getting information out to the consumer) or a "pull" system (allowing consumers to access a database and search for the information they need). A valuable dissemination tool that accommodates both the push and pull systems is the Joint Deployable Intelligence Support System (JDISS). Through this system, intelligence users have access to the Joint Worldwide Intelligence Communications System (JWICS), the sensitive compartmented information (SCI) portion of the Defense Information System Network (DISN). JDISS provides a means of dissemination of intelligence products, as well as a means of communication.

Another tool under development to help the timely and accurate dissemination of intelligence products is the joint intelligence virtual architecture (JIVA). The purpose of JIVA is to create intelligence products that are "modular" or "living," in the sense that they can be updated much more quickly than typical intelligence reporting procedures. JIVA gives intelligence analysts the ability to insert new information into intelligence products, and to update geographic coordinates or personalities. Through a process called collaborative white boarding (CWB), intelligence analysts can store information and reach decisions on intelligence assessments in a much more timely manner, thus improving the responsiveness of the intelligence cycle to the end user's needs. The drawback with JIVA is that it requires "buy-in" by all agencies within the intelligence community who share production responsibilities for certain products.

The final phase is evaluation and feedback. In the original intelligence cycle model, feedback and evaluation were understood, rather than articulated, because there has always been an evaluation process with HumInt reporting.[16] Intelligence analysts are routinely asked to respond to collectors who have cited a user's requirements in an intelligence information report, whether the information provided was "of major significance, of value, or not of value."[17] The intent of the evaluation process is to ensure that the collectors are responding to the users' needs and to realign collection efforts when they are not. When it falls short, the system must be adjusted to correct the deficiency or else the entire discipline suffers.[18]

The Intelligence Community

To further understand how the entire process works, a brief review is needed to focus on the makeup of the intelligence community. As mentioned in the last chapter, at the national level is the Director of Central Intelligence (DCI), who is also the head of the Central Intelligence Agency. The National Security Act of 1947 tasks the DCI with "directing and conducting all national and foreign intelligence and counterintelligence activities."[19] Intelligence production and collection activities by all thirteen of the U.S. intelligence agencies come under the purview of the DCI, including those belonging to the Department of Defense. See Diagram 2-1 for a graphic representation of the intelligence community.

The DIA is the national-level intelligence organization with responsibility for intelligence production and collection in support of DoD elements. DIA and other national-level intelligence agencies provide a wealth of collection platforms, resources, and capabilities that can be tasked by the DCI to support military operations. It is also the conduit through which joint commands get their time-sensitive requirements. In

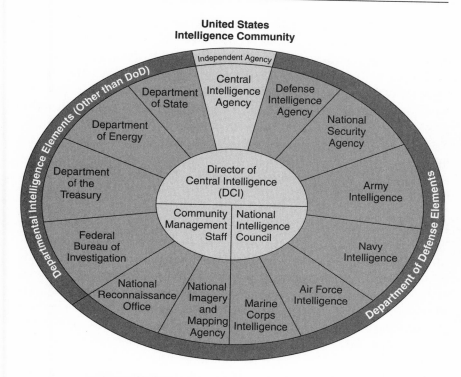

Diagram 2-1 The Intelligence Community

addition, the DIA, National Reconnaissance Office (NRO), CIA, and other intelligence agencies work with the Joint Staff J-2 to run the National Military Joint Intelligence Center (NMJIC) in the Pentagon. Liaison officers from these national agencies sit in the NMJIC and respond to time-sensitive requests from the field. They also provide the "reach back" to national HumInt, ImInt, and SigInt systems that can be tasked to support operational requirements.

In the early 1990s, following the Gulf War, Joint Intelligence Centers (JICs) were established at each combatant command with the intent to "improve the quality of intelligence support to the warfighter while decreasing the resources required to provide such support."[20] By "pushing" more collection and analytical responsibilities to the CCs, the DoD also created the linkages needed to ensure connectivity to the national-level agencies. In other words, along with the formation of JICs at the combatant CCs came liaison officers from the DIA, CIA, and NSA and the necessary command, control, communications, computers, and intelligence (C4I) linkages to these national-level agencies through the JWICS and the JDISS.

When a CC needs to form a joint task force (JTF) to support an operational requirement, he or she can tailor the intelligence support needed to perform the mission. This is accomplished with the formation of a National Intelligence Support Team comprised of representatives of the CIA, DIA, NSA, and/or others, tailored to support a JTF commander. An operational element of the JIC, called a Joint Intelligence Support Element (JISE), may also be formed to directly support the JTF. A JISE is usually formed during a crisis and serves under the JTF J-2 to "manage the intelligence collection, production, and dissemination for a joint force," and it may also function in a "split-base" mode, which allows the JTF J-2 to keep the bulk of the intelligence support at the JTF home base, with a smaller reach within the Joint Operational Area.[21]

In addition to deploying theater-level intelligence resources, a J-2 must also plan for the integration of service and coalition intelligence capabilities and resources. Each service component has a unique array of intelligence collection platforms that must be integrated with the operational collection plan. Coalition capabilities may or may not be available, given the constraints on intelligence sharing and releasability of intelligence methods and sources. The services and coalition partners also bring unique challenges concerning connectivity and the ability to disseminate intelligence. Standardization of equipment and the use of standard operating procedures will always be problems, yet increased use and availability of JWICS and JDISS are helping to alleviate this situation.

IO and Intelligence Preparation of the Battlespace

In order to respond to the unique needs of IO, the intelligence community has revised its support through the means of conducting intelligence preparation of the battlespace (IPB). This is an analytical process or method used by individual intelligence officers and their staffs as well as intelligence organizations. The ultimate goal of IPB is to reduce uncertainty and allow a commander to focus combat power to counter an adversary's most likely course of action. In other words, done correctly, IPB gives the intelligence officer a level of confidence in his or her assessment beyond "gut instinct."

In the joint community, the joint intelligence preparation of the battlespace is defined as:

> The analytical process used by joint intelligence organizations to produce intelligence assessments, estimates, and other intelligence products in support of the joint force commander's decision making process. . . . The process is used to analyze the air, land, sea, space, weather, electromagnetic

and *information environments,* as well as other dimensions of the battle-space, and to determine an adversary's capabilities to operate in each.[22]

The inclusion of "information environments" in the definition of battle-space indicates how the intelligence community has come to recognize the unique challenges posed by information operations and the unique intelligence requirements needed for conducting IO in peacetime, as well as IW in times of conflict.

Joint intelligence preparation of the battlespace is a four step process: (1) defining the total battlespace environment, (2) describing the battle-space's effects, (3) evaluating the adversary, and (4) determining and describing the adversary's potential courses of action, particularly the most likely courses of action, and the courses of action most dangerous to friendly forces and mission accomplishment.[23]

In step one, defining the battlespace environment, a number of planning factors must be considered. For example, what are the operational limits of the joint operational area? What is the joint force commander's mission? What are the unique characteristics and dimensions of the battle-space? What amount of detail is required? What information already exists in databases, and so on? What information requirements exist and what intelligence collection is needed? These types of questions occur early on in the planning and directing phase of the intelligence cycle. What is unique about intelligence support to an IO plan is that these questions, and others, must be asked far in advance of any operational deployment and do not lend themselves to crisis action-planning. Intelligence products, to be timely, accurate, and available for IO planning, often need to be "ware-housed" and easily retrievable. Also, due to the interagency structure of IO it may not necessarily be a DoD organization requesting the information.

Step two of the IPB process describes battlespace effects. Traditional considerations such as military aspects of terrain cannot be ignored, nor can the impact of weather, mission, troops available, time, and so on.[24] Yet in the information battlespace, other effects must also be considered:

- Media access and availability (foreign and domestic)
- Information systems usage (government, industry, military)
- Internet access (population)
- Critical infrastructures and architecture (power, banking, telecommunications, etc.)
- Public opinion (both domestic and foreign)
- Other actors present and their agendas (including nongovernmental organizations)

Diagram 2-2   Examples of IO Targets

Therefore, in an IO environment practically every facet of an operation (military, diplomatic, etc.) must be considered against the potential impact of all these factors.

In step three, evaluating the adversary, IO seeks to target the adversary decision-maker, as shown in Diagram 2-2. Therefore, identifying those who actually make the decisions and their closest advisors is critical. In developing the appropriate human factors analysis of these individuals, other questions need to be asked:

- What is their psychological mindset?
- What are their political goals and strategic objectives?
- What intelligence sources and methods do they employ and trust for their information?
- What are their biases?

These questions cannot be answered overnight, thus requiring long-term data collection, in particular a dependence on HumInt for understanding a decision-maker's intent. A recent development in the IC for this step of the IPB process is the use of Human Factors Analysis Centers (HFACs). Many joint operational commands are developing HFACs within their Joint Intelligence Centers (JICs) in order to conduct this detailed level of analysis. Considering that the ultimate target set of IO is the adversary decision-maker, such developments as the HFAC are a welcome addition to the intelligence community.

In the final step of the IPB process, the analyst must determine an adversary's course of action. Here, intelligence officers must take a position and argue their case before the commander on what they believe the adversary will or will not do. In the recent case of North Atlantic Treaty Organization (NATO) military operations in Kosovo, past indicators of Serbian President Milosevic's behavior led to the intelligence assessment that after a couple days of NATO bombing, Milosevic would give in to NATO demands and support the Ramboullet Accords.[25] Because he did not capitulate, NATO planners conducted a "plan as you go" strategy until the conflict ended months later. As will be shown later in this book, IO was not utilized correctly as an enabling strategy in Operation Noble Anvil.

The relationship between IO and the joint intelligence preparation of the battlespace process is a continuous interaction between the J-2 and J-3 when developing a joint campaign plan or theater engagement plan. It is also evident that intelligence support of IO works best in the deliberate planning process when there is significant lead time before an operation, rather than in a crisis action mode. That's not to say that intelligence cannot support short-fuse contingencies, but when it comes to supporting the unique requirements for conducting IO, the system is not as flexible or responsive as one might wish. JP 3-13 states that offensive IO requires intelligence support that is "broad-based" and "dedicated" with "significant lead time" and "early employment." It also states that "intelligence must be collected, stored, analyzed, and easily retrievable" and that intelligence collection should include "all possible sources."[26]

The entire concept of information superiority, as mentioned in Chapter 1, depends on intelligence superiority. In order to have a tighter decision loop than the adversary, one needs to know that adversary better than one ever had to in the past. Banking on technological improvements in information processing, collection platforms, and the entire reconnaissance, surveillance, and target acquisition process alone can produce vulnerabilities, particularly in the field of human intelligence. If the United States has learned anything from its intelligence failures of the past and the shortsightedness of its intelligence policy in the 1970s, it is that technical means of intelligence collection alone cannot confirm or deny an adversary's intent. What is needed is a long-term investment in recruiting the appropriate HumInt sources to develop the analytical tools required for human factors analysis (adversary decision-makers), and also investment in structuring appropriate counterintelligence assets (to counter an adversary's IO efforts, particularly denial and deception operations) so that intelligence will truly provide the support of information operations envisioned in JV 2020. Such a response will require increased national-level visibility for conducting IO and greater involvement of the interagency community in defining IO as a strategic priority that commands SecDef attention. One means of emphasizing the growing threats to U.S. critical infrastructures posed by cyberterrorists, state-sponsored information warfare, or other information threats is to categorize these in the NSS as being on a par with other threats, such as those posed by weapons of mass destruction. Only then will the appropriate intelligence sources and requirements be adequately funded and directed to provide the necessary support of information operations. As has been seen since the events of September 11th, this emphasis on increasing intelligence sharing between different government agencies is drastically changing the way that these organizations do business. For, as mentioned earlier, if IO is to be truly functional, not only does it need good intelligence support, but it also must be integrated across USG agencies and coalition nations as well. As will be noted in the following section, it may be some time before IO reaches its full potential because of these limitations on intelligence sharing among our allies.

The Releaseability Issues of IO

IO is still hampered by many issues, foremost among them is the classification and releaseability of information to coalition nations. To begin with, the CCs conduct most of the DoD IO planning in an IO cell. These small groups are part of the operations directorate and are tasked to ensure that IO issues are integrated into the larger plan. They coordinate with the J-39 from the Joint Staff as well as other IO agencies. Therefore, a problem arises with releaseability issues because much of IO planning is still conducted at high classification levels. Normally these plans are at least held at the "Secret United States Only No Foreigners" level and often rise well above the "Top Secret/Sensitive Compartmented Information" status. There are also special access programs (SAPs) and special technical operations (STO) that, though not exclusively devoted to IO, nevertheless play a significant role.

These high levels of classification for IO limit its appreciation by a wider audience. If you look at some recent operations to see how IO has been conducted, you will notice that there are very few unclassified "lessons learned" available. Because most of IO planning is still conducted in classified areas or STO cells and the use of SAPs has been high, there has not been a consensus to declassify much material in the last few years. Of course, the classification level also depends on which capability or related activity you are concerned. Much of the electronic warfare, destruction, psychological operations, and operations security principles are common enough among the different nations to allow dissemination at a "Secret" level on these issues. Likewise, public affairs and civil affairs are most often conducted at the unclassified level in order to gain the media attention and private contacts needed to conduct business. However, it is the computer network attack and deception capabilities that have most often been tightly held and that have tended to remain the most non-releaseable in reference to foreign disclosure to other nations. Computer attack and defense programs are some of the newest technology and perhaps the most far-reaching of the IO weapons. Therefore, their use has been highly restricted, not only to other nations, but even within the United States.

The basic document that provides guidance on the releaseability of intelligence to foreign nations by the United States is the National Disclosure Policy (NDP-1). In addition, other documents such as the NSS, Joint Publications (JPs), and DCI directives have all stated, in different fashions, their views on foreign disclosure. Most publications, including JV 2020, stress that the United States will operate in a multi-national or coalition environment, so it is crucial that intelligence be shared when possible. However, the true guidance is laid out in NDP-1. Even DoDD S3600.1, published in 1996, refers to NPD-1 for its disclosure policy. In the United States, the basic policy for disclosure of classified military information includes:

- Classified military information will be treated as a national security asset which may be shared with foreign governments and international organizations only when in the interest of the United States.

- Foreign governments must protect classified U.S. military information with a degree of security comparable to that received in the United States.

- Disclosures must always be consistent with U.S. foreign policy objectives.

- Disclosures will be made only when it can reasonably be assumed that information would not be used against U.S. interests.

Another factor to consider is which U.S. agency or command is the originator of the intelligence, for this determines who can downgrade or declassify the information. The other major factor is which countries are involved. If these nations are allied in a bilateral relationship with the United States or if the nations are currently involved in a coalition operation, they may be allowed access to certain releasable material. Thus, the basic principles for foreign disclosure listed in NDP-1 include:

- Intelligence sharing for common threat perception;
- Level of classification; and
- Dissemination architecture.

At an unclassified level, that is about all one can say concerning NDP-1 and IO. One cannot disclose what the rules and policies are for disclosure, nor can one go into great detail about what the different agencies in the U.S. intelligence communities are doing. Although there has been some work to downgrade this technology to a lower classification level, at the current time, much of that effort does not go below "Secret/No Foreign." Thus the use of IO, and in particular CNA, in a coalition environment is going to be constrained and is probably going to be rather closely held and secret for the foreseeable future. This will, of course, constrain multinational operations and perhaps limit the effectiveness of IO, but the amount of limitation will depend on how these technologies are treated in future doctrinal updates.

To make IO truly successful, you must have detailed and integrated planning. You must include as many players as necessary, reaching out beyond the traditional military agencies to the private sector as well as to allies and coalition partners. That takes trust all around; however, as we all realize, that trust will take time to develop and grow. IO is a relatively new warfare area that is still evolving and the releaseability of its capabilities and related activities is an issue that may not be resolved without a lot of effort by all parties concerned. To work successfully, the United States will therefore need to bring its allies into the fold and release more aspects of these classified warfare areas. However, the same is true for all coalition partners. To be effective, information must be shared on a timely basis. There is a concerted effort to downgrade many aspects of IO, but this effort may not get to a level that will satisfy many allied nations in the near future. The capabilities of portions of IO, primarily CNA, are such that it may be a while before these technologies are released on a more general level by the United States. But that is not to say that, in general, the level of interoperability among military forces concerning IO has not risen greatly over the last few years. As military commanders get more familiar with IO and its capabilities the interoperability and sharing issues will likely lessen between coalition partners.

Yet there are attempts to increase the information sharing between coalition partners. U.S. Joint Forces Command (JFCom) is the recognized authority within the DoD for joint task force (JTF) interoperability, and as such they developed a system for secure information exchange. In an attempt to replace the dubious "sneaker" net operations, JFCom built the Coalition Multi-Level Hexagon Prototype (CHMP). This system is composed of the following six functions that work together to ensure fast information exchange between allied nations in a secure and flexible manner:

- Marking standards
- Document marking
- Digital labels
- Personal authentication
- Hardware required for CHMP
- Security management

The real key to this whole system, besides the obvious administrative benefits, was the development of a CHMP Hexcard to allow for personal authentication. Similar to an ATM card, this system stores an individual's fingerprint, clearance levels, citizenship, and level of access. When inserted into the workstations and servers, the Hexcards allow for the proper transfer of data between allied nations. These attempts may not solve all the problems of information releaseability, but at least they are a step in the right direction.

Summary

The role of intelligence is crucial to the implementation of IO by the United States. It must be horizontally integrated across the strategic, operational, and tactical levels. This will hopefully ensure that the key pieces of intelligence are delivered to the decision-maker in a timely manner. Thus the role of the senior intelligence officer (J-2) on a joint staff cannot be overstated. The J-2 must understand the needs of the operational community and work hand-in-hand with the J-3 to ensure that the commander is properly supported. As one CC has commented, "IO can't wait" and the J-2/J-3 must be fully integrated into the concept of operations at all levels of planning and executing IO.

CHAPTER 3

Information Protection

The Challenge to Modern Bureaucracies

"The best defense is a good offense."

Anonymous

Military operations have traditionally been categorized as either offensive or defensive in nature. In addition, weapons systems have also been described in this vein, such as air defense artillery or interceptor aircraft. Yet, any defensive operation also has an inherently offensive side when it neutralizes the adversary's intent, either through active measures, such as physical destruction, or by using a set of passive computer firewalls. Thus, the idea that a weapon is totally defensive or that one can truly protect one's self with purely defensive measures is usually not a concept fully accepted by U.S. military personnel. As a general rule, most military operations have both offensive and defensive components, and this is evidenced in doctrine as well. The joint information operations doctrine refers to both offense and defense as the two primary subdivisions.[1] For the purpose of this text, this categorization will be used; however, the very nature of IO and its relation to agencies other than the military makes this distinction less critical and probably will prove to be a disservice in the end. The overall trend is toward greater integration, or as stated in JP 3-13, "fully integrated offensive and defensive components of Information Operations are essential."[2]

Defensive Information Operations

Defensive IO is concerned with more than just computer-based information systems, yet you might not know that from reading recent publications. There has been a tremendous amount of press on CND and IA, yet as you learned in Chapter 1, defensive IO is much more than that.

That's because information is critical to all military operations, traveling over a number of mediums including satellites, broadcast media (television and radio), facsimiles, cellular phones, and so on. Each of these systems therefore possesses its own vulnerabilities that can be exploited by an adversary, and these are not always computer-based. For example, the television footage of dead U.S. Army Rangers being dragged through the streets of Mogadishu, Somalia in 1993 did much more damage to U.S. political resolve and ultimately impacted warfighting capability more than a few well-placed precision-guided munitions could have. Likewise, today, through the use of low-cost commercial or over-the-counter technologies, any adversary can significantly impact the most sophisticated military organization's information processing and decision-making capabilities. Former Secretary of Defense and current Vice-President Dick Cheney once commented that advanced technologies have made third-class powers into first-class threats.[3]

So what exactly are defensive information operations? The definition found in JP 3-13 describes defensive IO this way:

> The integration and coordination of policies and procedures, operations, personnel, and technology to *protect and defend* information and information systems. Defensive information operations are conducted through information assurance, OpSec, physical security, counter-deception, counter-propaganda, counterintelligence, electronic warfare, and special information operations. Defensive information operations ensure timely, accurate, and relevant information access while denying adversaries the opportunity to exploit friendly information and information systems for their own purposes.[4]

Interestingly, the only change in this definition that differed from the preliminary draft of JP 3-13 is the italicized phrases. Are these significant? Adding OpSec, hardly—probably an oversight of the first draft. But, adding "protect and defend" implies more active rather than passive measures. To provide an illustration, when one installs an intrusion detection system on one's home, he or she is employing passive protection measures. When the same individual keeps a loaded 9 mm pistol in his or her nightstand drawer, he or she is employing a more active defensive measure to protect the family and property. Such an analogy is applicable to information operations, where the DoD employs more active defensive measures to protect information systems, rather than relying simply on intrusion detection systems,[5] as shown in Diagram 3-1.

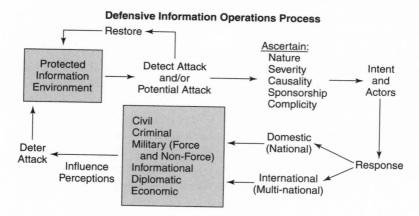

Diagram 3-1 Defensive IO Process

Within the discipline of defensive information operations, there are two main goals:

- To minimize friendly IO system vulnerabilities to adversary efforts.
- To minimize friendly mutual interference during the operational employment of IO elements and capabilities.

A good example of the first goal is emphasized by some lessons learned during the war in Kosovo. During that air campaign, a reporter from *US Today*, the national newspaper, was allowed to fly a bombing mission in a B-52 out of Barksdale AFB in Louisiana. He reported the sights and sounds of the mission and proudly upheld the tradition of emphasizing the hometowns and families of the bomber's aircrew. Within days, supporters of the Serbian government were harassing and threatening this aircrew's families using information gained from this front-page story. Hence in Operation Enduring Freedom, during the fall of 2002, the news media only used the first names or call signs of military personnel that they were interviewing. This is a distinct change in official policy. Likewise, the second part of the equation is that one needs to protect oneself from oneself. Information fratricide is a very real concern on the information battlespace, and the deconfliction and the use of the electromagnetic spectrum becomes even more complex and necessary as available bandwidth decreases.

JP 3-13 describes defensive information operations as a process. Initially proposed by the Joint Staff J-6K as the defensive IO model, it has been incorporated into the doctrine by J-39 for the conduct of IO. The diagram contains four embedded processes. The first is to *protect* one's protected information environment. The process as shown in Diagram 3-2, involves identifying which information systems are the most critical to an organization and determining the appropriate policies, procedures, technologies, and operations necessary to safeguard these critical systems. A critical part of the protect process involves IA, which is discussed later in this chapter. This process seeks to ensure that the information getting into one's system is just as valid as the information that is going out.

The second process identified in the diagram is *detect*. In other words, how do individuals know that their organization is under an information attack? Many DoD elements are working to develop a system of indications and warnings specifically designed to detect if and when an information attack may occur. What makes this process so difficult is the need for a fully integrated indications and warnings capability across the spectrum of IO, which involves agencies other than just the DoD. As was

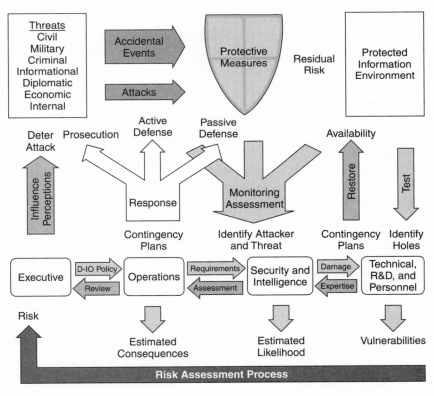

Diagram 3-2   Risk Assessment Process

explained in Chapter 1, a fully integrated national infrastructure protection plan is crucial for linking DoD with other federal agencies, as well as industry, in order to ensure the appropriate level of intelligence and warning sharing occurs in a timely matter.

The third process is *restore*. The key in any cyberattack is to maintain transparency of operations and not let on that the intruder is having an impact on the operations of one's organization. In addition, it is also very important to ensure redundancy in the restoration process, namely, to ensure that there are appropriate back-ups and routing capabilities to reduce the impact of an intrusion. Here again, cooperation between the government, society, and industry is crucial for ensuring continuity of operations and reducing vulnerabilities. A good example is a fully integrated system of CERTs that can share timely information on intrusions and sources of the attacks, information that is crucial for ensuring continuity of operations in any environment.

The final step in the defensive information operations process is *respond*. The linkage back to the other processes is necessary in order to determine the nature of the attack—its severity, sponsorship, and so on—and to further identify the actors and their intent. The response DoD takes will ultimately hinge on these criteria, as well as whether the attack has domestic or international implications. Depending on the source and the intent, the DoD may not be involved in responding to an attack, particularly if there are domestic law-enforcement equities at stake. If the military is involved in responding to an information attack, there will likely be very restrictive rules of engagement enforced and the traditional military reaction (steel on target or overwhelming forces) may not be the most appropriate reaction.

Currently U.S. military commands are implementing a series of defensive IO measures aimed at increasing their knowledge of and response to information attacks. Exercises such as Eligible Receiver and Warrior Flag seek to educate and train commands about the dangers of information system vulnerabilities and to help them develop the appropriate safeguards and response mechanisms. In March 1999, the Chairman of the Joint Chiefs of Staff published a directive on prescribed InfoCons in order to standardize the reporting and declaring procedures for these conditions. These new directives were quickly assimilated into the military structure and are now considered a standard part of any base infrastructure. As mentioned earlier, the JTF-CNO and StratCom have the responsibility for promulgating changes to these procedures on a daily basis, and overall, the InfoCon process has worked relatively well over the last few years.

These procedures mentioned here, and some of the technologies alluded to earlier, are good steps in building a defensive posture, but the most important component of any plan is people. No amount of high-tech intrusion detection devices or command-directed policy can overcome the effectiveness of unit personnel who take an active interest and role in the defense of their command. Yet all too often, leadership is too enamored with new protection devices or software, when instead their money would be better spent on adequately training their people. As we have tried to emphasize throughout this book, IO is much more than just computers, and it truly is the personnel aspect that makes this new warfare so effective.

In summary, defensive information operations include a range of capabilities aimed at reducing vulnerabilities to information and information systems. The weak link in all of these procedures, as mentioned earlier, is the personnel, the primary systems operators. The largest threat today comes from the trusted insider; however, just as significant a threat also comes from the individual who is untrained, unprepared, and uneducated in operating very technical components of a command's information system. By oversight or error, serious damage can occur by even the most well-intentioned individuals, crippling a military organization's "eyes and ears." In an age when information is an element of national power and decision loops need to become tighter, an educated and technologically sophisticated military workforce is more crucial than ever.

Information Assurance and Computer Network Defense

As stated in the first chapter, it is the application of new technology that has revolutionized warfare by taking the elements of power and dispersing them to the people. The computer has been the primary driver of this huge change, and though IO is more than just computers, it is in this area that a huge deal of current emphasis on funding and research are currently being conducted. So it is only natural that with a growing awareness among government and military officials to the vulnerability of their networks and information systems, more money and more emphasis will be placed on IA and CND, as shown in Diagram 3-3. Fortunately for these officials, the development of new doctrine has facilitated this emphasis on defense. In 1996, the JCS released a white paper that formulated the direction and future strategy of U.S. military forces. Entitled *Joint Vision 2010* (JV 2010), this policy document had many features, but one key area was its dependence on information superiority (IS). It was considered crucial to future military operations that the United States achieves IS, which was further subdivided into three areas. IO is therefore a subset of IS, which in turn is a subset of JV 2010. Remember that relationship, because it has changed as policy has matured.

Information Assurance

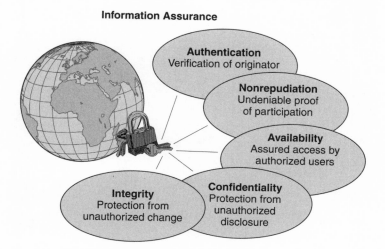

Diagram 3-3   Information Assurance

As mentioned earlier, IO does not address either IA or CND directly, but within the military community they have been incorporated under the greater IO umbrella. Some of that philosophy may go back to the old C2W doctrine which had offensive and defensive components. Whatever the case, you can draw a direct connection between IA and JV 2010. Information assurance is an umbrella term, which covers many portions of the information protection construct, or the defensive IO area, including CND.

The armed forces of the United States increasingly rely on critical digital electronic information and communications capabilities to store, process, and move and visualize essential data in planning, directing, coordinating, and executing military operations. The implementation of the complex, massive Global Information Grid (GIG) through implementation of combat digitization and network centric warfare concepts using advanced communications and computer technologies means that warfighters are becoming so technologically dependent that system failure or disruption can completely change the tempo of the battle.

In this broad threat environment, where every connection to a network must be regarded as a potential avenue of attack, IO must defend not only our own information and information systems, but also must affect adversary information and information systems to suppress its capability to be utilized against us, as shown in Diagram 3-4. Mentioned briefly before, this is done primarily through IA, which is a major subset of IO. Information assurance supports the full-dimensional protection aspect of JV 2010/2020

Information and Information Systems Vulnerabilities

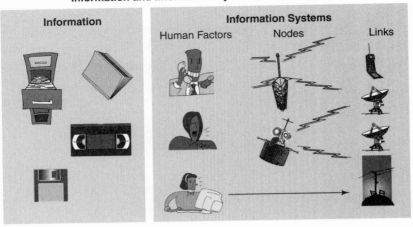

Diagram 3-4 Information and Information Systems Vulnerabilities

and comprises actions at the tactical, operational, and strategic levels that protect and defend information and information systems by ensuring *availability, integrity, authentication, confidentiality,* and *non-repudiation.* It also includes providing for restoration of information systems by incorporating protection, detection, and reaction capabilities.[6]

To do this, IA seeks to ensure the security of information in its myriad of forms, not just information transferred using telecommunications or stored computers. Thus there is a close, synergistic relationship between counterintelligence, operations security, communications security, information security, and information systems security, all of which seek to protect information from hostile access and exploitation. This concept is much more expansive in scope than classic information systems security which many people normally tend to relate to. IA is also much broader than CND, which has gotten a great deal of attention recently. It encompasses those communications and computer network management functions that seek to provide for continued operations in the event of accident, natural disaster, deliberate act, and adverse operational environment, as shown in Diagram 3-5.

The concept of IA also covers what was previously referred to as peacetime defensive IW and facilitates operational coordination with agencies outside of the DoD, including the civil agencies of the federal government, industry, and the public sector, as shown in Diagram 3-6. It should not be confused with defensive IO, which is more of an active/

Growing Threats to Information and Information Systems

Diagram 3-5 Growing Threats to Information and Information Systems

Information Environment Protection

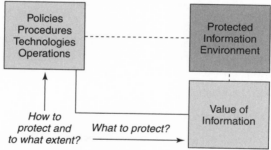

Diagram 3-6 Information Environment Protection

reactive process that addresses the manifestation of specific threats. The defensive IO process is cyclic and underpinned by the use of risk management as shown earlier. This change in terminology also addressed difficulties for those non-DoD agencies for which the idea of being involved in *warfare* has caused political difficulties through preconceptions of conflict. Thus the adoption of the term IA has also reflected wider recognition that issues such as Critical Infrastructure Protection (CIP) are much larger than the DoD and that successful IO, particularly IA, relies on the whole of government and community conducting an integrated approach and cooperation of military, intelligence, government, industry, and public efforts. However, with respect to military operations, IA is a critical operational readiness and warfighting issue that commanders need to be cognizant of at all times, and in particular factor into their battlespace appreciation when going into combat.

This holistic approach to IO protection taken by IA is good, because since the demise of the Cold War, the threats to the United States have changed. Threats come from a variety of sources, including natural physical elements and forces that have been influenced or used through human intervention—accidental or unstructured activities through to hostile, structured, and deliberate misuse or attack. However, recent reports indicate that threats from internal sources have not significantly diminished from previous years and the figures from the FBI/Computer Security Institute survey of 1998 are still valid. Organizations surveyed indicated that 89 percent of their perceived threat is still internal.[7] This emphasis on protecting the computer systems and information is primarily covered in the discipline of computer network defense. In this context, CND refers to those operational IA and defensive IO measures implemented in the computer network environment to defeat both internal and external threats, as shown in Diagram 3-7. It takes IA out of the administrative sphere and places it in the operations community with the other elements of IO. Starting in the year 2000, there has been a move to combine CNA and CND into an operational discipline known as CNO, but that is still undergoing analysis.

This emphasis on CNA has produced a tremendous amount of interest in the vulnerability of the United States and its military forces. In effort to validate anecdotal evidence of the possible impact of CNA on military

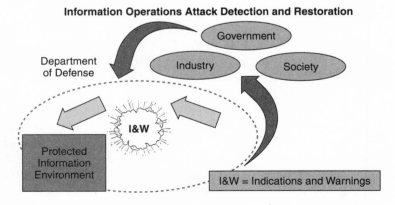

Information Operations Attack Detection and Restoration

Diagram 3-7   IO Attack, Detection and Restoration

operations and the National Information Infrastructure (NII), the JCS sponsored a series of exercises known as *Eligible Receiver*. The most famous of these was mentioned earlier and was called *ER '97*, which demonstrated in June 1997 that hostile forces could penetrate national infrastructures and DoD networks using CNA and other techniques to adversely affect the government's ability to conduct military operations. A number of non-DoD agencies were also involved including the FBI, DoJ, Department of Transportation, DoS, CIA, NRO, and the NSC. The threat consisted of an NSA Red Team replicating the threat from a domestically situated but state-sponsored team operating on behalf of a nation that had refused direct military confrontation with the United States. This nation concluded that the United States was now so militarily and economically dependent on vulnerable information systems that a non-attributable CNA operation offered a viable option. The aim of these attacks was to alter U.S. policy and delay or deny its ability to respond militarily to avoid detection and arrest.

The Red Team, using only open-source intelligence and hacker tools available on the Internet, was able to fully demonstrate the vulnerability of DoD and national-level system and network vulnerabilities. The rules of engagement allowed the team to conduct actual attacks on DoD systems and conduct simulated attacks on NII systems. Lessons learned from *ER '97* emphasized the need for effective vulnerability assessments, network indications and warnings, appropriate command and control, a designated cyberdefense command, consequence management, and interdepartmental/interagency planning, procedures, and processes. Probably the most important lesson learned from *ER '97* was the need for a central DoD agency to be in charge. As mentioned earlier, DISA is normally responsible for protecting the NII; however, in reality, because it is a combat support agency, DISA cannot order a CC or government agency to change any policies. It eventually took over eighteen months to solve this problem, and it was only completed with the formation of the JTF-CND.

In the meantime, during February and March of 1998, the U.S. military, government, and research and development sites experienced a large number of systematic intrusions that were determined to be related to one another. Code-named *Solar Sunrise*, the timing of these activities was very suspicious because it coincided with another build-up of U.S. military personnel in the Middle East in response to tensions with Iraq over United Nations weapons inspections. The intruders penetrated many unclassified U.S. military computer systems, including air force bases and navy installations, DoE national laboratories, NASA sites, and university sites. The timing of the intrusions, and the apparent origination of some activity from the Middle East, led many government officials to suspect that this could be an instance of Iraqi CNA aimed at disrupting the U.S. military build-up in the region. Subsequent investigation and detailed research by the NIPC and the FBI, working closely with Israeli law enforcement authorities, determined after several days that two juveniles in Cloverdale, California, and an individual with several accomplices in Israel were the perpetrators. Once again, the need for a central government agency to coordinate an appropriate response was needed, but not available yet. *Solar Sunrise* showed the need for DoD to constantly coordinate with law enforcement agencies, especially the FBI, when dealing with unidentified computer intruders.

If *Eligible Receiver '97* and *Solar Sunrise* were eye opening operations to the vulnerability of U.S. systems, then the next series of incidents codenamed *Moonlight Maze* was the true wake-up call. Dr. John Hamre, the Deputy Secretary of Defense at that time, described the *Moonlight Maze* events to a congressional committee in 1999, stating bluntly: "We are in the middle of a cyber war." The *Moonlight Maze* event was enormous and officials have publicly stated that the intruders systematically accessed and exploited hundreds of unclassified but sensitive computer networks used by the DoD, DoE, NASA, various defense contractors, and several universities. A large amount of technical data related to defense research was copied and transferred to Russia. One defense technician trying to track the computer intruder is said to have watched in amazement as a document from a naval facility was "hijacked" from a print queue to a location in Moscow right in front of him. The first *Moonlight Maze* attack was detected in March 1999. Three months later, U.S. agencies were able to monitor a series of intrusions as they occurred and traced them back to seven dial-up Internet connections located near Moscow. The FBI attempted to determine if the United States was subject to intelligence collection over the Internet conducted by Moscow's prestigious Russian Academy of Sciences. Intense attacks continued until at least May 1999, and the FBI investigation remains open. *Moonlight Maze* has been the most insidious and focused assault yet on sensitive DoD and government computer networks. Yet Russia may just be the last line in a long series of transactions of a very determined data-mining effort. Therefore, caution is required with respect to attributing *Moonlight Maze* to the Russian government, as there has been no definitive evidence of a military or intelligence connection. "It could turn out to be Russian organized crime," stated one source, who also indicated, ". . . and they could be acting as a front for the intelligence community."[8]

These were just the beginning of what seems to be a never-ending series of computer attacks on the USG. Most of these incidents come in the form of malicious software, which may be a virus or worm. These programs are designed to replicate and spread themselves within a network or server. Generally this is done without the operator's knowledge, and in many cases it can severely compromise the machine's ability to operate. Prudent system administrators insist on installing and updating the latest anti-virus software such as those programs sold by Symantec, McAfee, or IBM. Probably the first virus that served as a wakeup call to the average citizen and the USG was the Melissa virus. On March 26, 1999, this new virus first appeared and by March 30th, four days later, it had successfully infected over 70,000 e-mails. This was the first virus to prompt a warning to be issued by the FBI, and for many Americans it was their first exposure to the dangers of a virus. The majority of viruses are file viruses, but the Melissa version was different. File viruses infect other files by attaching themselves as an executable file, with an .exe or .com extension, indicating executable program code. These are exactly the kind of files that a firewall looks for and tries to exclude from a network. Likewise, the other common virus is a boot sector or partition table virus that is normally hidden until an operator executes the boot-up process.

Melissa was different, however. Melissa was a macro virus, or really a worm, which means that its infectious code is part of a macro; in this case a Microsoft Word application. Therefore it was able to bypass most firewall and virus-scanning programs because it was not an .exe or .com extension. Once Melissa was resident inside an operator's computer, it proved especially deadly. Because it moved as an e-mail, once opened by the unsuspecting operator the macro quickly read the first fifty names from the Microsoft Outlook address book and then forwarded itself on to them as another e-mail with an attachment. By spreading so quickly, the Melissa virus overloaded the servers and eventually crashed a number of them in a classic denial of service attack.

What was also different about the Melissa virus was that lessons had been learned over the prior eighteen months in the wake of *ER '97,* and so some new organizations were in place to effectively combat the virus. As was emphasized already, both the JTF-CND and NIPC were established and were able to quickly assess the threat, develop a defensive strategy, and coordinate a whole host of defensive actions. For the first time in DoD and USG history, someone could answer the question about who was in charge of network defense.

Solar Sunrise and *Moonlight Maze,* and to an extent the Melissa virus, also validated and reinforced the findings of *ER '97* by clearly demonstrating the need for a cooperative approach by federal and DoD organizations to nationally manage the defensive IO battle. Possibly the most important lesson learned from *ER '97, Solar Sunrise,* and *Moonlight Maze* has been the requirement for an effective and efficient incident and vulnerability reporting system. Comprehensive incident reporting is critical to determine whether an intrusion is local and isolated or part of a more widespread activity. This reporting system must accept data from both automated systems, such as intrusion detection systems and firewall logs, as well as manual incident reports based on personal observation and analysis of users and system administrators. Streamlining the incident reporting and analysis process requires standardization of reporting formats, handling procedures, data transfer and maintenance procedures, and integration with response capabilities at all levels.

To meet these requirements the DoD implemented a four-tier incident and vulnerability reporting structure with reporting and analysis at global, regional, service, and local levels, as shown in Diagram 3-8. All local military network operations and security centers (NOSCs), whether deployed or at standing bases, camps, posts, and stations, report upward through either or both of the two functional/command chains, one from a network perspective, the other from an operational perspective. Thus the process of reporting through DISA regional NOSCs, many of which are collocated with warfighting CCs, is consistent with the traditional network management processes for reporting network problems. From a command and operational impact perspective, reporting is conducted through individual service or regional CERTs, and reflects more traditional operational reporting. Both of these levels report to the DISA GNOSC and the co-located

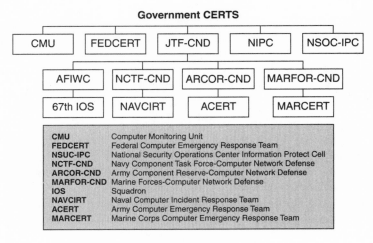

Diagram 3-8 DoD CERT Architecture

JTF-CNO. So, if nothing else has been learned from these three major computer attacks, the need for a centralized reporting and dissemination system has been recognized for its importance.

In addition, another major lesson learned during ER '97, Solar Sunrise, and Moonlight Maze was that in order to defend DoD computer networks properly, a commander that was responsive to higher authority was needed. Therefore, as mentioned earlier, to answer the deficiencies from these cyber incidents, the JTF-CND was formed on December 30, 1998, to provide a single command with operational authority to coordinate and direct the defense of the DoD computer systems and networks. Once established, JTF-CND became the command that could direct CCs and service units to operationally change settings and set InfoCons to protect against the threat. Originally reporting directly to the secretary of defense, JTF-CND has since become a command that reports directly to the Space-Com as of October 1, 1999, when that CC officially took over the mission of computer network defense for the whole of the DoD. Likewise, as mentioned earlier, one year later on October 1, 2000, SpaceCom also assumed responsibility for the complementary offensive CNA role. The merger of these two missions came on April 1, 2001, when the JTF-CND was renamed the JTF-CNO. A final change occurred on January 10, 2003, when the JTF-CNO was transferred to StratCom.

A further development in the incident-handling process that was also mentioned earlier is the development of InfoCons, which support the threat warning of computer network-based activities. This system provides a structured, coordinated approach consistent with Defense Threat Conditions (DefCons), with graduated responses to defend against CNA. InfoCon measures obviously focus on implementation of computer network-based protective measures to meet a changing threat level. Each level reflects a defensive posture based on the risk of military operational impact through the disruption of friendly communications and information systems. Countermeasures at each level include preventive actions such as changing passwords, actions taken during an attack such as enabling all system logging, and damage control/mitigating actions such as physical disconnection from the network. InfoCon levels are as follows:

- NORMAL (normal activity)
- ALPHA (increased risk of attack)
- BRAVO (specific risk of attack)
- CHARLIE (limited attack)
- DELTA (general attack)[9]

These descend in order of increasing defensive protection, and, once set, all commands must abide by the restrictions delineated in the Chairman's Memorandum CM-510-99.

To date the InfoCon system has received a lot of publicity, but there are still some concerns that it may not be addressing all of the needs of the commands in the IA arena. Therefore in June 2000, SpaceCom hosted a conference in Colorado Springs to look at revising the original document. From this meeting came a plan to concentrate on the following areas:

- Commander's assessment criteria
- Directed actions
- Operational reporting

Since that time, revisions have been made to the basic document, and these were implemented during fiscal year 2001.

As mentioned earlier, prior to the formation of the JTF-CNO, no single DoD entity had the authority to coordinate and direct a department-wide response to a network attack. Together with the multi-agency NIPC, the JTF-CNO now forms a strong collaborative team for dealing with attacks on DoD systems/networks and the wider NII. At the global level, the GNOSC reports to and coordinates with the NSA National Security Operating Center/Information Protect Cell (NSOC/IPC) and NIPC, as shown in Diagram 3-9. The JTF-CNO liaisons with the Joint Staff and the National Military Command Center to provide the operational analysis and recommendations concerning CNA and CND issues to the CJCS.[10]

Other major changes conducted over the last three years include the addition of the IA Vulnerability Alert (IAVA) process and vulnerability assessments that give the information-protection managers more tools to make decisions to better assess the information threat. IAVA is the comprehensive distribution process for notifying appropriate agencies about system vulnerability alerts and work around/countermeasure information. It has developed into an extremely formal process that requires acknowledgment of the receipt of the vulnerability alert by the different commands to the reporting authority. The system also has time requirements that require a specific response from the recipient of the alert confirming that they have implemented appropriate countermeasures. Network vulnerability assessments, sometimes referred to as *on-line surveys* or vulnerability assessments, use technical scanning software as an active method of validating the installation and configuration of GIG information systems and whether they meet appropriate requirements. In addition, another major growth area is the use of IA Red Teams to conduct active vulnerability

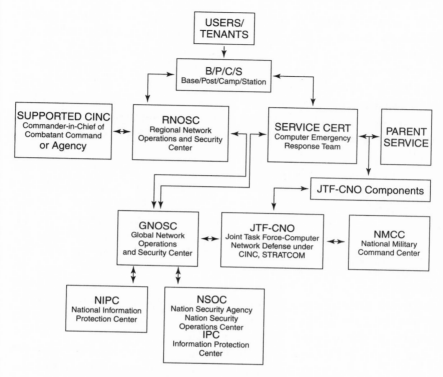

Diagram 3-9   DoD CERT Coordination

assessments of networks replicating the threat posed by computer intruders. All of the U.S. military services maintain information warfare centers
whose capabilities have been employed mainly under exercise conditions.
Red Teams provide the ability to validate not only the configuration of the
systems, but also to test procedures and systems user's/administrator's ability to detect and react to CNA, something that an on-line survey cannot do.

All of the measures previously listed are part of an overall USG defensive strategy called *defense in-depth* (DiD). This approach integrates the
capabilities of people, operations, and technology to achieve strong, effective, multi-layer, and multi-dimensional protection. DiD attempts to ensure
that the level of protection of one system is not undermined by the vulnerabilities of other interconnected systems, ascribing a minimum standard of
assurance that all systems connecting to the environment must attain. The
idea is that the successive layers of defense will cause an adversary who pen-
etrates or breaks down a barrier to promptly encounter another DiD barrier,
and another, thereby increasing the likelihood of detection and offering the
opportunity for one of the defensive mechanisms to defeat the attack. To be

effective, the DiD strategy must protect against a variety of attack methods. To do this, a corresponding variety of complementary defensive mechanisms must be employed so that the weaknesses of one barrier are offset by the strengths of another.

The three key components of a comprehensive DiD strategy are people, technology, and operations. Appropriately trained and certified people, operating in accordance with well-defined policies and procedures, using certified and accredited networks and systems with layered and distributed security technologies, are the key to DiD. Security background investigations, clearances, credentials, and badges for critical network personnel are required given the internal threat. To implement this strategy, on July 14, 2000, the Deputy Secretary of Defense issued a memorandum on the implementation of the recommendations of the IA and Information Technology (IT) integrated process team on training, certification, and personnel management in the DoD.[11] This memorandum provided a framework for training and certification of IA personnel and directed the service chiefs, CCs, and DoD agencies to address retention issues given the increasing "brain drain" from DoD to private industry.

For the technical portion of DiD strategy, a number of policies and instructions have been recently promulgated. Ensuring that systems components are certified and accredited in accordance with DoD Instruction 5200.40, the "Defense Information Technology Certification and Accreditation Process" (DITSCAP), throughout the system's life cycle is another critical procedural aspect of the DiD process.[12] As with any other military activity, policy drives IA operations by establishing goals, actions, procedures, and standards. The ASD/C3I issued two new departmental policies with respect to IA, specifically IA 6-8500, *Guidance and Policy for Department of Defense Global Information Grid Information Assurance,* and IA 6-8510, *Global Information Grid Information Assurance Implementation Guidance.* These documents detail the implementation of DiD and require the development of three types of doctrinal policy. In addition, during the spring and summer of 2002, the OSD also began to issue a series of IA and CND-related DoD instructions. The overarching publication was a new series of DoDD 8500 instructions which superseded all of the old IA documents, including DoD Directive 5200.28, DoD Manual 5200.28-M, and DoD Chief Information Officer (CIO) Memorandum 6-8510 No. 8500. This directive specifically applies to:

- All DoD-owned or controlled electronic information systems and technologies that receive, process, store, display, or transmit DoD information, regardless of classification, mission category, or sensitivity. This includes portable computing devices such as laptops, handhelds, and PDAs operating in both wired and wireless modes.

- This policy does not apply to information systems to which DCI Directive 6/3, that is, Sensitive Compartmented Information and special access programs for intelligence apply under the purview of the Director of Central Intelligence.

Key highlights of this new DoD policy include:

- Information assurance shall be a visible element of IT-dependent and IT-related investment portfolios, and shall be reviewed and managed relative to the return on investment, contributions to mission outcomes, and contributions toward the achievement of strategic goals and objectives in accordance with Division E of the Clinger Cohen Act of 1996.

- The DoD shall organize, plan, train for, and conduct the defense of DoD computer networks as integrated CND operations that are coordinated across multiple disciplines in accordance with DoDD O-8530-1.

- For systems requiring logon authentication, the minimum requirement shall be a properly administered and protected case-sensitive password consisting of at least eight characters. The password shall be a mix of upper-case letters, lower-case letters, numbers, and special characters, including at least one of each (e.g., Passwd2!).

- The use of public key infrastructure (PKI) certificates and biometrics for positive access control shall be in accordance with published DoD policy and procedures. These technologies shall be incorporated in all new acquisitions and upgrades wherever possible.

- DoD information systems shall regulate remote access and access to the Internet by employing positive technical controls such as proxy services and screened subnets, also called de-militarized zones (DMZ), or that are isolated from all other DoD systems through physical means.

- All DoD information systems shall be certified and accredited in accordance with the DoD DITSCAP, DoD Instruction 5200.40.

- Connection to the DISN shall comply with established connection approval procedures and processes.

- Use of public key services in DoD information systems shall be in accordance with published DoD public key policy and associated guidance.

- All IA-related government-off-the-shelf (GOTS) and commercial-off-the-shelf (COTS) hardware, firmware, software components, and IT products shall be evaluated and acquired in accordance with National

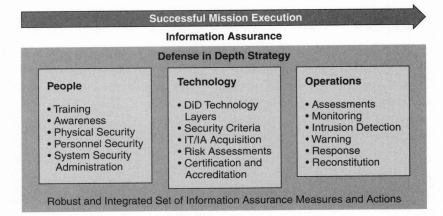

Diagram 3-10   IA DiD Strategy

Security Telecommunications Information System Security Policy (NSTISSP) No. 11 and other applicable national and DoD policy.

- Mobile code technologies shall be categorized and controlled to reduce their threat to DoD information systems in accordance with DoD component policy and guidance.
- A Designated Approving Authority (DAA), an Information System Security Manager (ISSM), and an Information Systems Security Officer (ISSO) shall be appointed for each information system and enclave operating within or on behalf of the DoD.

In addition, the J-6K division of the Joint Staff has also proposed a new IA publication that will provide the overarching operational guidance for IA in the joint context.

That the DoD has adopted DiD, as outlined by the JCS J-6 Directorate, is a time-honored tactic to combat not only malicious threats but the accidents and acts of nature as well, as shown in Diagram 3-10. The basic idea is that you construct a series of defensive layers to protect your networks and systems against attack. Using a medieval castle as an analogy, J-6 and in particular J-6K has outlined a comprehensive series of protective layers that must be defended:

- Network and infrastructure
- Enclave boundary
- Computing environment
- Supporting infrastructures

Technical Security Measures	Availability	Confidentiality	Integrity	Authentification	Non-Repudiation
Cryptography		X	X	X	X
Password, PIN, Biometrics				X	X
Digital Signatures				X	X
Firewall				X	
IDS		X	X	X	X
Virus Scanner	X	X	X		
Vulnerability Checker	X	X	X	X	
Guard		X			
Proxy Server		X			
System Monitoring Tools	X		X		
TRANSEC	X	X	X		
TEMPEST	X	X			
Anti-Tamper	X	X	X	X	
PDS	X	X	X		
Multiple Paths	X		X		
Backup	X		X		X

Diagram 3-11 IA Technical Security Measures

All of this is spelled out in the DoD Guidance and Policy for Information Assurance. Together with this policy, the JCS has also been working closely with the OSD staff, in particular the DIAP, to develop and procure equipment to protect each one of these layers.

DiD relies on this layered approach to ensure that there is not just one critical node that can be defeated. To do this, advances in technology have helped the manager and system administrator build protection into their networks as shown in Diagram 3-11. Some examples of these new trends include:

- High-speed networking (ATM/100 Mbps Ethernet)
- Wireless technology
- DNS Domain Name Security

Notwithstanding the need for advanced technology to help network security, there are already products on the market that can help a company, command, or agency raise its network standards. These include:

- Public key infrastructure (PKI)
- Virtual private network (VPN)

- Firewalls
- Intrusion detection systems (IDS)
- Virus scanners
- Smart cards
- Secure applications
- Honeypots and honeynets
- Complex attack plan generators

There is a clear trend towards providing commercial tools and products which integrate and manage these individual products. A simple secure sign on (S3) application, which provides a central process allowing the user to be authenticated and then to access all of the systems for which they are authorized , uses multiple items of the previous list. However, the convenience and efficiency of S3 in accessing information for decision-making must be balanced against the very critical need to secure the information and systems involved. The S3 system represents a single point of failure in the network it serves.

PKI is a large part of the federal strategy for providing security throughout the government. Based on the distribution and control of public and private keys, PKI will have three layers of assurance. The DoD has implemented a high-level PKI system called FORTEZZA that will operate in a similar manner. No matter what the particular PKI system is called, they all operate in generally the same manner.

Likewise VPN is an attempt to secure the consumer with secure communications and data protection while still allowing access to limited-public network access. To date there are three types of VPN products loosely based in the following categories:

- Hardware-based systems
- Integrated systems
- Software-based systems

Firewalls, on the other hand, are mechanisms to deny or allow access to a particular network. They are great devices for limiting access to unauthorized connections—but as was shown by the Melissa virus in 1999, they have their limits. Because many of the older versions were looking primarily for executable or .exe files, this particular virus slipped by many firewalls because it was a Word macro. The generation of firewalls currently being developed includes the stateful multi-layer inspection (SMLI) and SOCKS protocols. Offered by companies such as IBM, DEC, and

CyberGuard, these third-generation applications offer good protection against the current level of threat to a network. Basically, SMLI works by analyzing the entire data stream, at all levels. Because most applications come with significant extra code, SMLI will examine each packet and compare it to a database. SOCKS, on the other hand, inspects both incoming and outgoing traffic at the session layer, applying security on a packet-by-packet basis.

Another technical security measure included in this area is IDSs, which are a natural complement to firewalls. Whereas the firewall will attempt to filter malicious and subversive activity, IDSs monitor and try to detect attempts to subvert security measures already in place, similar to a burglar alarm. Think of the firewall as a locked door that keeps unwanted visitors out and think of the IDS as your security guard who watches for suspicious activity. As mentioned earlier, IDSs monitor not only outside threats, but more importantly they also monitor the insider as well. In today's environment, the greatest threat to a network is not from the teenage hacker but instead from the disgruntled employee. At the present time, IDSs can be broken down into two categories: host and networked-based. The former normally protects one workstation or server while the latter are designed to work outside the perimeter firewall and record all activities of a particular network.

Access and authentication are huge issues with respect to computer security and IA. Strong authentication of user identity uses several different and separate factors: something the user has, something the user knows, and something the user is. Passwords are an example of something the user knows. However, the days of just using a password to protect your system are long gone. There are many reasons why passwords alone don't work, but the greatest weakness is that it is humans that use them. Because a password is something that must be remembered, many operators use a password that can be easily guessed, say their birthday or child's name. Others make the mistake of using the same password on all their accounts, rendering them vulnerable. Likewise, many write down their passwords on yellow stickies and then leave them where anyone can see them. For those reasons and more, many systems and networks are using biometrics to identify a person, vice a password. However you will notice that in the most recent DoD instruction, mentioned earlier in this chapter, the details of a proper password are delineated in exact detail, to make passwords stronger and less likely to be broken.

Biometrics is another attempt to control access to networks and computer systems. Biometrics identify something the user is. Biometrics is the science of measuring the human body and, believe it or not, there are many different parts of your body that are unique and can be measured to identify you and distinguish you from someone else. There are a number of methods for conducting biometrics with respect to identifying a single operator. Many companies are producing systems or kits that you can use, and the Biometrics Consortium is trying to develop standards for biometrics in this rapidly expanding field.[13] In addition, the USG is also investigating a number of these processes, and some have already been incorporated within the DoD. Following are a few of the current biometric processes:

- Face recognition
- Fingerprint recognition
- Speaker recognition
- Hand geometry/finger geometry
- Iris recognition
- Dynamic/static handwritten signature
- Retina scan
- Keystroke dynamics
- Wrist vein

Smart cards are another attempt to use technology to prevent unauthorized use. Smart cards represent something the user has. Based on the PKI and FORTEZZA-based systems mentioned earlier, smart cards allow access while still meeting the authentication, confidentiality, and integrity requirements of IA. These cards have embedded microchips that support different operating systems and various secret key encryption algorithms. They also support PKI infrastructures, as mentioned earlier, namely with the common access card (CAC) technology. Starting in 2000, many DoD personnel were issued CACs to replace their traditional military identification cards, and eventually all service-members will utilize this technology. Current smart card use is estimated to be 100 billion transactions a year with much more growth projected for the future.

Honeypots are systems that are configured to draw the initial attention of attackers, delaying their progress and allowing time to analyze the virus, worm, or other mode of attack. Honeynets serve much the same purpose, but present a subnet seeming to be an interesting target but in actuality being a collection of trapping and analysis tools. In addition, complex attack plan generators are a response to the rapidly increasing complexity of networks and speed of network attacks. Due to the multiple vulnerabilities of network assets and multiple modes of exploiting those vulnerabilities by multiple threat agents, anticipating rapid attacks that have multiple steps before such an attack is mounted is vital to a well-planned defense. Plausible complex attack plans can be generated based on known (or hypothetical) vulnerabilities and network architectures. The plans may contain forty or more steps, far more than usually encountered currently. Different plans can be evaluated using simulated or real networks to assess the impact of the attack. This technology can be applied as a defensive and an offensive tool. Its effective use depends on clearly understanding how complex systems work and accurately knowing the state of a given network and its architecture.

However, as emphasized elsewhere in this book, to conduct a proper cyberdefense program, the DoD must utilize personnel and policy as well as technology, tools, and products. Thus, the third portion of the DiD strategy is operations, which contains many different programs. The Defense-Wide Information Assurance Program (DIAP), as mentioned in Chapter 1, is a management process and structure established to centralize IA efforts within DoD. The program was designed to integrate and coordinate DoD IA activities to:

- Provide a structure to monitor and coordinate IA readiness
- Establish IA responsibilities and authorities across the department

The DIAP is the chief mechanism allowing the DoD chief information officer to ensure IA information technology and resources are effectively managed and implemented to meet DoD operational requirements. In addition, to help assess computer applications, the NIAP framework was established by a 1997 agreement between the Commerce Department's NIST and the NSA to evaluate new technology. All of these organizations are designed to help the operator and were covered in detail in the first chapter.

A View of Defensive Information Operations

The threat to information structures is wide and asymmetric; therefore, the need for appropriate IA has never been greater. Insiders are still the greatest threat, despite the increasing level of threat from external sources, requiring appropriate physical and personnel security and counterintelligence measures. As noted earlier, IA is not composed of magic black box solutions but rather of an integrated suite of capabilities encompassing people, training, technology, and policies.

> [T]he amorphous nature of today's security environment means the threats will be far more difficult to anticipate and counter. These asymmetric threats pose end games that are still potentially devastating to countries and alliances. We must, individually and collectively, anticipate these types of threats and have the courage to deal with them.[14]

To conclude, JV 2010/2020 is vulnerable without supporting IA, and the computer network threat to civilian infrastructure supporting operations must be considered and adequately planned for. JV 2010/2020 will not be protected by any single organization, plan, or policy, and the distribution of stakeholders and resource owners should convince any skeptic that the task is broad and needs to be coordinated at the organizational, national, and international levels. Commanders must accept that everyone has a stake in IA and that the capabilities required to effectively protect U.S. and allied information infrastructure will only be achieved through cooperative, collaborative efforts across the domain of critical infrastructure supporting military operations.

Counterterrorism Information Operations

During the Cold War between the United States and the Soviet Union, American policy centered on the possibility of a nuclear attack. This threat was well known and efforts were conducted to prevent it at any cost. However, since the collapse of the bipolar world power paradigm in 1991, predictability has broken down and the United States is now involved in crisis engagement around the world. Often these crises set the U.S. military forces against the "rogue of the month" instead of a single force or even an ideology. Hollywood portrays terrorists in many forms with various causes, from the recent James Bond film *The World is Not Enough*, where petroleum distribution was paramount, to *Air Force One*, where the freedom of foreign political prisoners was the terrorists' cause. These films may be fictional in story line, but these threats and others like them may be all too real for U.S. national security.

This new reality became shockingly close to Americans when visions of terrorist acts were flashed across millions of television sets as office workers climbed their way out of the World Trade Center rubble in February 1993. A little over two years later, the scene was replayed when terrorism struck the nation's heartland on the morning of April 19, 1995 at the Alfred P. Murrah Federal Building in Oklahoma City. These incidents have greatly raised the awareness of terrorism within the United States. In this section, anti-terrorism and counterterrorism information operations (CTIO) will be explained. In addition, the culture of counterterrorism will be analyzed to provide a foundation to better understand the reasons why governments respond to terrorism in the manner that they do. Finally, the importance of PDD-62 is discussed to demonstrate the changing use of information operations management.

As much as these earlier incidents did to heighten the awareness of terrorism, nothing could of course compare or change American attitudes more than the attacks of September 11, 2001. The effects of those four airplane crashes and the loss of over 3,000 lives has forever altered any ideas citizens of the United States had about CTIO, and although the outcome of the retaliation operations are still ongoing, it is an understatement to say that the United States will never be the same.

What Is Terrorism?

According to Bruce Hoffman in his latest work, *Inside Terrorism*, "Like 'Internet'—another grossly overused term that has similarly become an indispensable part of the argot of the late twentieth century—most people have a vague idea or impression of what terrorism is, but lack a more precise, concrete and truly explanatory definition."[15] As Walter Laqueur writes, "No definition of terrorism can possibly cover all the varieties of terrorism that have appeared throughout history."[16] Thus, the interpretation of the classification of terrorism is most important in the development of a definition of terrorism.

Although some scholars have cited up to 109 different definitions of terrorism, the FBI currently uses the following definition: "The unlawful use of force or violence against persons or property to intimidate or coerce a government, the civilian population, or any segment thereof, in furtherance of political or social objectives."[17] The FBI also classifies acts of terrorism as either domestic or international, depending on the origin, base, and objectives of the terrorist organization. For the purposes of this book, the following sub-definitions are also used:

- *Domestic terrorism*: The unlawful use, or threatened use, of force or violence by a group or individual based and operating entirely within the United States or Puerto Rico without foreign direction and whose acts are directed at elements of the USG or its population, in the furtherance of political or social goals.

- *International terrorism*: The unlawful use of force or violence commit-
ted by a group or individual, having some connection to a foreign power
or whose activities transcend national boundaries, against persons or
property to intimidate or coerce a government, the civilian population,
or any segment thereof, in furtherance of political or social objectives.

Combating Terrorism

Inherently, terrorism is an attack on the legitimacy of the established
order, a negation of that system.[18] The Clinton administration divided
combative terrorism efforts into two major categories:

- *Counterterrorism,* which includes *offensive* methods to combat terrorism (e.g., efforts to preempt and prosecute terrorists).

- *Anti-terrorism,* which includes *defensive* methods to combat terrorism
(e.g., protection against and management of the consequences of an
attack).

In order to keep terminology simple, both counterterrorism and anti-
terrorism are combined together and labeled as counterterrorism in this
book, in order to explain the overall U.S. strategy to combat terrorism.
Therefore, "the goal of counter-terrorism is to prevent and combat its
use."[19] Thus the methods to prevent terrorist acts are inherently based on
IO due to the psychological aspects that even the *possibility* of a terrorist
attack inflicts. Good examples of this are seen in the trauma experienced
by Americans in the immediate aftermath of the events of September 11,
2001, as false alarms continued to be broadcast by both the media and the
government. It seems that both sides, terrorists and policy makers, have a
political purpose and the primary audience is a third party or the body
politic.[20] In order to conduct successful counterterrorism information
operations against terrorist groups and to deter anti-American attacks, the
credibility of the United States is absolutely crucial. "Policymakers must
above all demonstrate competence . . . the government may not always
win, but it must show that U.S. policymakers, not terrorists, are in
charge."[21] Many analysts credit President Bush with bringing calm to the
nation with his speech on September 20, 2001 demonstrating that the
United States would fight back against the al Qaeda terrorists.

Contrast the mood in the United States after this historic discourse
with the Carter administration's dismal foreign policy failures involving
the Iranian hostage crisis. The capture of fifty-five Americans and the
inexcusable duration of their internment had a vastly negative psychological impact on the American public. Since that horrible debacle, the United

States has implemented retaliatory strategies against terrorists and their sympathizers that rely on U.S. credibility in terms of international diplomacy and the use of military force. These strategies include air strikes in 1986 on Libya in retribution for the terrorist bombings, and the strikes against Sudan and Afghanistan in 1998 as retaliation for the bombings of the United States in Africa. Secretary of State Madeleine Albright addressed critics of this course of action at an American Legion Convention on September 9, 1998:

> Some suggest that by striking back, we risk more bombings in retaliation. Unfortunately, risks are present either way. Firmness provides no guarantees, but it is far less dangerous than allowing the belief that Americans can be assaulted with impunity. And as President Clinton has said, our people are not expendable.[22]

So what is the official U.S. foreign policy against terrorism? It starts with a foundation that establishes that military forces will be employed where necessary and appropriate to prevent and punish terrorist attacks with four policy tenets that symbolize counterterrorism rhetoric:

- Make no concessions to terrorists and strike no deals.
- Bring terrorists to justice for their crimes.
- Isolate and apply pressure on states that sponsor terrorism to force them to change their behavior.
- Bolster the counterterrorist capabilities of those countries that work with the United States and require assistance.

If you look back over the events of Operation Enduring Freedom, and the wide-ranging operations that are currently being conducted by the White House, DoD, DoS, DoC, and Treasury in order to combat al Qaeda, you can see that these doctrinal statements are still being adhered to.

Fundamentals of CTIO

Counterterrorism information operations (CTIO) builds on this verbal strategy and targets seven types of terror groups:

- State supported
- Social revolutionary left
- Right wing
- National-separatist

- Religious fundamentalist
- New religions
- Criminal[23]

IO becomes a factor because, like all humans, these terrorists are subject to misinformation campaigns, perceptions management operations, and "coercive diplomacy," that is, the threat of force as an influence on behavior, which requires the enemy to believe that force can be and will be used.[24] It is these activities that the United States employs to accomplish its military and political objectives—specifically, a misinformation campaign intends to persuade perspective members from joining a group or influencing public opinion concerning a terrorist organization. For example, a perceptions management operation consisting of information manipulation similar to the Electronic Disturbance Theater acts described in the introduction can be used. Furthermore, the enormous growth of NGOs is a crucial tenet of any public perceptions management operation, and these organizations need to be addressed in any IO planning scenario. But perception management operations can also backfire, as they did to the United States in early October 1993, when horrific images of Army Rangers' bodies being dragged through the streets of Mogadishu were shown around the world. Finally, the United States can also use "coercive diplomacy" to portray itself to be the defender of law and order to the body politic, both domestically and internationally through the United Nations, by condemning those states that support terrorism. International treaties or conventions or the actual funding for counter-terrorism operations also accomplishes this "confidence building" approach with its allies.

PDD-62 *Counter-Terrorism*
These events raised awareness of the threat posed by terrorism to the U.S., but tangible policy outcomes took a little longer to emerge. The first key Clinton Administration response to the evolving terrorist threat was to promulgate Presidential Decision Directive (PDD) 39—*US Policy on Counter-Terrorism,* dated 21 June 1995.[25] PDD-39 articulated a four-point strategy that sought to reduce vulnerability to terrorist acts, to deter terrorism, to respond to terrorist acts when they occur, and measures to deny terrorists access to weapons of mass destruction (WMD). The four-point strategy integrated both domestic and international measures to combat terrorism. PDD-39 was novel in that it specifically identified the vulnerability of CNI and potential terrorist employment of WMD as

issues for concern. But in general, PD-39 generally lacked sufficient bureaucratic teeth to achieve meaningful outcomes.

But a clear outcome of PDD-39 was to raise the profile of CIP policy. U.S. policy CIP was not a novelty per se. But previously CIP policy had tended to be overshadowed by other elements of U.S. national security policy.[26] And the increasing inter-connectedness of CNI had created a range of dependencies and vulnerabilities that were historically unprecedented. PDD-39 directed the Attorney-General to establish a committee to review and report upon the vulnerability of CNI to terrorism.[27] In turn, this committee—the Critical Infrastructure Working Group (CIWG)—noted that CNI was vulnerable not only to attack by physical means, but conceivably also by computer-based means. As the result of the lack of knowledge of the cyber threat, the CIWG recommended that a presidential commission be established to more fully investigate this matter.

The CIWG's recommendation led to the establishment of the Presidential Commission on Critical Infrastructure Protection (PCCIP) on 15 July 1996 by Executive Order (EO) 13010.[28] Whilst the PCCIP was primarily a response to PDD-39, in an informal sense it also consolidated a range of uncoordinated CIP policy development activities occurring across government.[29] EO 13010 also directed the establishment of an interim Infrastructure Protection Task Force (IPTF) within the Department of Justice (DOJ), chaired by the Federal Bureau of Investigation (FBI).[30] The purpose of the IPTF was to facilitate coordination of existing CIP efforts whilst the PCCIP undertook its work. The IPTF was chaired by the FBI so that it could draw upon the resources of the Computer Investigations and Infrastructure Threat Assessment Center (CITAC), which had been set up there in 1996. In effect, the IPTF represented the first clear effort to establish coordinating arrangements across different government agencies and with the private sector for CIP.

The final report of the PCCIP—*Critical Foundations*—was released in October 1997.[31] The key finding of the PCCIP was that whilst there was no immediate overwhelming threat to CNI there was need for action, particularly with respect to NII protection. The PCCIP recommended a national CIP plan; clarification of legal and regulatory issues that might arise out of such a plan; and a greater overall level of public-private cooperation for CIP. From late 1997 to early 1998, the PCCIP report underwent interagency review to determine the Administration's policy response.[32] But in February 1998 concrete outcomes were already beginning to emerge, as the interim IPTF was amalgamated with the CITAC, made permanent and renamed the National Infrastructure Protection Center (NIPC).[33]

The recommendations of the PCCIP were given practical expression on 22 May 1998 with the release of two policy documents: PDD-62—*Combating Terrorism,* and PDD-63—*Critical Infrastructure Protection.*[34] These two documents were the culmination of the Clinton Administration's efforts at policy development for CT and CIP. PDD-62 was a direct successor to PDD-39. It provided a more defined structure for CT operations, and presented a focused effort to weave the core competencies of several agencies into a comprehensive program.[35] Also in common with PDD-39, PDD-62 sought to integrate the domestic and international elements of U.S. CT policy into a coherent whole.

The use of misinformation campaigns, perceptions management operations, and "coercive diplomacy" rely on a systematic planning approach to either operate independently or as a combined information operation. One of the first attempts to accomplish this was by the Clinton administration, which signed PDD-62 *Counter-Terrorism* on May 22, 1998. It was a much more systematic approach to combating the threat of terrorism than was previously directed in PDD-39, an earlier version of the counterterrorism doctrine.[36] This new doctrine provided a more defined structure for counterterrorism operations as well as the tools necessary to meet the challenges posed by terrorists who may resort to the use of weapons of mass destruction.[37] Furthermore, PDD-62 also served as a focused attempt to weave the core competencies of several agencies into a comprehensive program.[38] Specifically, this new counterterrorism doctrine clarified the specific roles of the many federal agencies responsible for counterterrorism, "from apprehension and prosecution of terrorists to increasing transportation security, enhancing response capabilities and protecting the computer-based systems that lie at the heart of America's economy."[39]

PDD-62 also established the Office of the National Coordinator for Security, Infrastructure Protection and Counter-Terrorism. It was intended that the Office would facilitate inter-agency activities, and in order to achieve this objective it established three senior executive working groups: the CT Security Group, the Critical Infrastructure Coordination Group and the WMD Preparedness Group. Collectively, the purpose of these working groups was to enhance capability for CT response. But the Office did not have the authority to mandate procedures to executive agencies, so its ability to affect change was limited.

Formerly occupied by Richard Clarke, who later was in charge of cybersecurity within the Office of Homeland Security, the national coordinator originally supervised a broad variety of relevant polices and programs including such areas as counterterrorism, protection of critical infrastructure, preparedness, and consequence management for weapons of mass destruction. He worked within the NSC, reporting to the president through the assistant to the president for national security affairs. Each year the coordinator produced an annual security preparedness report, and he also regularly provided advice regarding budgets for counterterror programs and led the development of guidelines necessary for crisis management.

As originally written, the Office of the National Coordinator for Security, Infrastructure Protection, and Counter-Terrorism acted as a facilitator and created three senior management groups that operated in the last two years of the Clinton administration:

- The Counter-Terrorism Security Group
- The Critical Infrastructure Coordination Group
- The Weapons of Mass Destruction Preparedness Group

Each of these unit's mission was to improve the methods of response to terrorist attacks, as shown in Diagram 3-12. However, the national coordinator did not have the authority to mandate procedures to the FBI, FEMA, or any other agency, therefore these groups were somewhat hamstrung in their ability to order changes in the government bureaucracy. However, these organizations were able to conduct exercises with other agencies during that time period, which helped train and upgrade interagency response.

PDD-63 was the document that implemented the recommendations of the PCCIP report, as interpreted through the prism of the inter-agency review panel.[40] PDD-63 identified twelve sectors of CNI, appointed

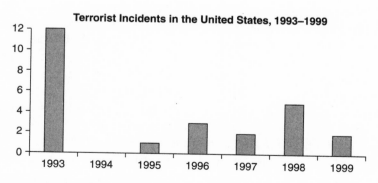

Diagram 3-12   Terrorist Incidents in the United States, 1993–1999. *Source:* FBI

government lead agencies for each of these sectors, and established coordination mechanisms for the implementation of CIP measures across the public/private divide. In particular, PDD-63 vested principle responsibility for the coordinating activities in the Office of the National Coordinator (which had been set up under PDD-62). PDD-63 also established the high level National Infrastructure Assurance Council (NIAC), to advise the President on enhancing the public/private partnership for CIP. PDD-63 also called for a National Infrastructure Assurance Plan (NAIP), which would mesh together individual sector plans into a national framework. Finally, PDD-63 authorized increased resources for the NIPC, and approved the establishment of sector Information Sharing and Analysis Centers (ISACs) to act as partners to the NIPC.

In the last year of the Clinton Administration, there were some minor changes to CT and CIP policies. Version 1.0 of a *National Plan for Information Systems Protection* was released in January 2000.[41] This was the direct result of the call in PDD-63 for a NAIP.[42] But reflecting the priority given to cyber security issues by the PCCIP, it primarily addressed NII protection rather than CIP as whole.

A good example of this new cooperation with respect to CTIO was the creation of the NIPC at the FBI's headquarters. As mentioned in Chapter 1, this organization is the latest example of the government's interagency effort. The center is a joint government and private sector partnership that includes representatives from the relevant agencies of federal, state, and local government, as well as from the private sector. H.R. 4210 the Terrorism Preparedness Act of 2000 was proposed at this time to improve coordination among federal agencies. This act established the Office of Terrorism Preparedness within the executive office of the president to coordinate and make more effective federal efforts to assist state and local emergency and response personnel in preparation for domestic terrorist attacks.[43] These organizations include the FBI, DoJ, FEMA, DoD, DoE, Environmental Protection Agency (EPA), the Bureau of Alcohol, Tobacco, and Firearms (ATF), Secret Service, Office of the Vice-President, offices under the Director of Central Intelligence (DCI), and the Department of Health and Human Services (HHS). A forerunner of the new Department of Homeland Security, the Office of Terrorism Preparedness had various sub-departments that coordinated counterterrorism and information operations issues across a variety of government agencies. However at that time, before the attacks of September 11, 2001, critics argued that the growth of federal programs to combat terrorism was excessive when compared to the relatively low number of domestic attacks, as shown in Diagram 3-12.

When the G. W. Bush Administration came to power in early 2001, there was some consolidation of existing CT and CIP arrangements. The collection of senior CT and CIP groups were consolidated into one Counter-Terrorism and National Preparedness Policy Coordination Committee (PCC) reporting to the National Security Council (NSC).[44] And whilst some debate occurred on future directions for CT and CIP policy, these bore no fruit prior to 11 September 2001. So in practice, during the first nine months of the G. W. Bush Administration, the bulk of the CT and CIP arrangements in place in the U.S. were largely a legacy of the previous Clinton Administration.

Across the period leading up to the 11 September 2001 attacks, the international aspect of the terrorist threat to the U.S. was becoming more evident. Incidents which demonstrated the international character of the terrorist threat included the 1993 WTC bombing, the June 1996 attack on the Khobar Towers complex in Saudi Arabia, plans to attack U.S. airliners in Southeast Asia in 1996, the attacks on U.S. embassies in Kenya and Tanzania, and the attack on the USS Cole in October 2000. In response to these incidents, both PDD-39 and PDD-62 incorporated measures to combat terrorism abroad. But whilst the international dimension of the evolving terrorist threat was acknowledged in policy, this was largely overshadowed by the domestic aspects of U.S. CT and CIP policies.

The terrorist attacks on 11 September 2001 led to fundamental changes to the U.S. government's approach to CT and CIP issues, as shown in Diagram 3-13. On 8 October 2001, EO 13228 established the Office of Homeland Security (OHS), to be headed by the Advisor to the President for Homeland Security (Tom Ridge, previously the Governor of New Jersey).[45] The purpose of the OHS was to develop and coordinate a national strategy to protect the U.S. against terrorist attack, in light of new threat posed by global terrorism. EO 13228 also established a high level Homeland Security Council (HSC), which was responsible for advising the President on all aspects of homeland security.

The following day, appointments were made for the National Director for Combating Terrorism (General Wayne Downing) and the Special Advisor to the President for Cyberspace Security (Richard Clarke).[46] Significantly, Downing had previously been the Commander-in-Chief of the U.S. Special Operations Command (USSOCOM), so his appointment reflected a greater prominence for the international (and overtly military) dimension of U.S. CT policy. Clarke's duties were formally spelled out on 16 October 2001, with the release of EO 13231. EO 13231 established the President's Critical Infrastructure Protection Board (PCIPB), which was to recommend policies and strategies for the protection of critical

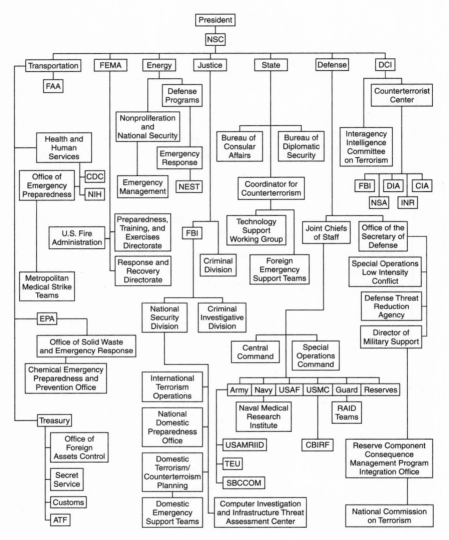

Diagram 3-13   Pre-September 11th CTIO
Organizational Architecture

information systems. The same EO also established the high level National Infrastructure Advisory Council (NIAC) to provide advice to the President on the same matter.[47]

In July 2002, the OHS released the National Strategy for Homeland Security.[48] The purpose of the strategy was to integrate all government efforts for the protection of the nation against terrorist attacks of all kinds. In effect, the strategy updated the measures enacted under PDD-63 in light of the post 11 September 2001 environment. The strategy did not create any new organizations, but assumed that a Department of Homeland Security (DHS) would be established in the near future. In September 2002, the PCIPB released for comment the draft National Strategy to Secure Cyberspace.[49] In effect, this document was the proposed successor to the Clinton Administration National Plan for Information Systems Protection. But whilst the issue of the draft plan was welcomed, concerns were expressed that it lacked the regulatory teeth to prompt action by the private sector.[50]

The most obvious consequence of the revised U.S. approach to CT and CIP occurred in November 2002, with the creation of the DHS.[51] This consolidated the bulk of U.S. federal government agencies dealing with homeland security (consisting of over 170,000 employees) into one department headed by a cabinet-level official. This represented the most fundamental change to U.S. national security arrangements since their inception in 1947. The DHS is comprised of five directorates (Management, Science and Technology, Information Analysis and Infrastructure Protection, Border and Transportation Security, and Emergency Response and Preparedness).[52] Significantly, the DHS closely resembled some of the measures that had been proposed by the U.S. Commission on National Security/21st Century (the Hart-Rudman Commission).[53] But it was only after the events of 11 September 2001 that the political imperative for significant organizational change for CT and CIP emerged.

Further action continued into 2003, with the release of three policy documents: the final version of the *National Strategy to Secure Cyberspace, the National Strategy for the Physical Protection of Critical Infrastructures and Key Assets* and the *National Strategy for Combating Terrorism*.[54] At the same time, EO 13286 abolished the PCIPB and the position of Special Adviser on Cyberspace Security. The NIAC was retained, but now reported to the President via the DHS. Combined with the departure of key staff associated with cyber-security issues, these measures raised concerns that cyber-security issues were being marginalized in the new arrangements. But overall, it is as yet still too early to assess how effective the new measures to protect CNI will be.

Overall, arguably the most evident aspect of the new approach to CT policy has been the involvement of the U.S. in substantial military campaigns in Afghanistan and (controversially) Iraq as part of the Global War on Terror.[55] But these two military campaigns have tended to overshadow a range of lower-key military and diplomatic activities across the globe.[56]

Also, it should be borne in mind that the GWOT does not represent a fundamental departure from the international aspects of CT policy articulated previously in PDD-39 and PDD-62, both of which incorporated a variety of measures to combat terrorism overseas. Rather, the difference between the GWOT and these previous policies is one of emphasis: the changed nature of global terrorism mandated a CT policy with a more evident international dimension.

Domestic Counterterrorism Operations

In the United States, the FBI is the lead agency for crisis management in the event of a domestic terrorist attack and FEMA is the lead agency for consequence management when a threat has subsided and the task is to restore order and deliver emergency assistance. This organizational arrangement was confirmed after the attacks of September 11th. Other agencies also contribute when necessary as shown in the Diagram 3-14. These federal agencies also conduct simulated exercises to test plus validate policies and procedures, as well as to determine the effectiveness of response capabilities to eventually increase the confidence and skill level

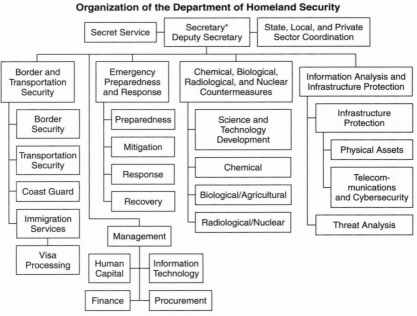

Organization of the Department of Homeland Security

* Legal/Congressional/Public Affairs included in office of the Secretary.

Diagram 3-14   Post-September 11th CTIO
Organizational Architecture

of personnel.[57] Furthermore, these tests are often conducted in a very public manner, which in essence becomes an IO perceptions management operation directed at the American people. This is done to reassure the body politic that the government is doing a great deal of preparation against terrorist attacks, but more importantly, it provides the impression that the government is doing something about terrorism, as shown in Diagrams 3-15 and 3-16. It is a verbal strategy as discussed before that demonstrates to the American people as well as to the rest of the world that the United States is engaged against terrorism just as the United States

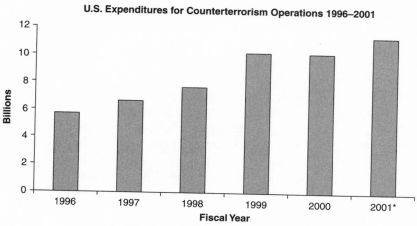

Diagram 3-15 U.S. Expenditures for Counterterrorism Operations, 1996–2001. *Source:* Office of Management and Budget (OMB)

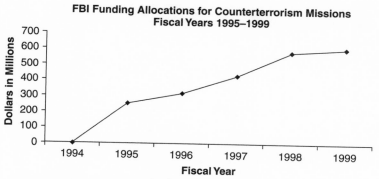

Diagram 3-16 FBI Funding for Counterterrorism, 1995–1999. *Source:* GAO

was engaged against communism. To give you an example of this, in May 2000, federal agencies conducted a ten-day response exercise designated TOPOFF, which stands for Top Officials. This operation brought together over 150 state and local emergency response planners and practitioners from across the nation to identify the objectives used to design the $3.5 million TOPOFF exercise. Agencies that participated in the exercise included the FBI, DoJ, FEMA, DoE, DoD, HHS, EPA, Department of Agriculture, the Department of Transportation, and the General Services Administration. To further test federal response, agencies performed National Capital Region 2000 (NCR-2000) in the District of Columbia and Prince George's County, Maryland. NCR-2000 tested the response to weapons of mass destruction events. Both events brought together a wide array of federal agencies to coordinate and practice a response to terrorism. These tests were successful; however there were still incidents where command and control issues plagued responders. Follow-up series of TOPOFF exercises have continued to occur, with the most recent being held during the summer of 2003 in Washington, D.C.

The bottom line is that counterterrorism has been a major mission for the federal government for a number of years, both in democratic and republican administrations. Even before the attacks of September 11th, spending for counterterrorism operations had nearly doubled in the last five years, and the budget for the lead federal agency for combating terrorism, the FBI, has also grown exponentially to prepare against the possibility of terrorist attacks. This growth in spending demonstrates that even before these barbaric acts counterterrorism was a top priority for the United States.

Partnerships for Counterterrorism: "Track Two Diplomacy"

Yet, U.S. counterterrorism policy is not formulated solely from reactive exercises and organizational changes in federal agencies. In fact, over the past several decades, there has been growing evidence that unofficial actors, including NGOs, are playing an increasingly important role in the development and implementation of government policies. To describe the efforts of ordinary citizens and unofficial organizations that resolve conflict, former U.S. diplomat Joseph Montville coined the term "track two diplomacy" in 1981.[58] The basic notion behind track two diplomacy is that government alone cannot achieve peace and conflict resolution. Unofficial, informal behind-the-scenes contact, either with policy recommendation publications or by attending multinational conferences, plays a vital role in conflict resolution and in promoting regional security.[59] NGOs also play an increasingly important role in combating international

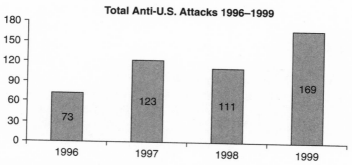

Diagram 3-17 Total Anti-U.S. Attacks, 1996–1999.
Source: Department of State

terrorism. These organizations not only act to influence policy decisions, they are also able to place accountability to combat terrorism on national and international organizations, like the United Nations Security Council. NGOs have the ability to sway public opinion or, more practically, to offer congressional testimony. To date, the most important function of these groups is their ability to help foster multi-national consensus concerning terrorism, thus forming an international front against terrorism, as shown in Diagram 3-17.

International Cooperation

No nation is an island, even when it is the sole remaining hegemony in the world. Foreign policy cannot be made around the world without balancing other potential great powers; it is therefore like a delicate dance—a ballet among different nations. If the United States were to operate as an imperialistic power that acts in a vacuum, it would quickly provoke other nations to challenge those policies. Therefore, in order to avoid such competition among nations, American foreign policy has often resorted to coalitions to conduct operations abroad. This approach seems to work best, especially when, as the sole super-power, the United States allows allies to initiate the strategy and to take the lead, thereby not giving the appearance of dominating the world. In this manner, the reigning hegemony can operate in the manner in which it desires and manage a coherent strategy without always fighting its critics.

This approach is consistent to American policy on terrorism. "International terrorism threatens U.S. foreign and domestic security and compromises a broad range of U.S. foreign policy goals," according to

Raphael Perl of the Congressional Research Service's Foreign Affairs and National Defense Division. He also notes that "Terrorism erodes international stability—a major foreign and economic policy objective for the United States."[60] Therefore, it is in the best interest of U.S. national security to seek cooperation from our allies to combat terrorism. This policy was recognized even before the attacks of September 11, 2001, and has been confirmed even more since that day. In this post-Cold War era, the United States relies on its allies to ensure global and regional security by using existing partnerships and multi-national institutions, such as the United Nations. Though there have been relatively few cases of domestic terrorism in the last four years, attacks against the United States have increased drastically. Data is not available for the last few years, but between 1996–1999, the number of total anti-U.S. attacks ranged from 73 to 169, an increase of 132 percent in four years.[61] One of those prominent attacks was the terrorists' strikes on the U.S. embassies in Nairobi, Kenya and Dar Es Salaam, Tanzania on August 9, 1998. Following these tragic disasters, Congressman Frank Wolf introduced the National Commission on Terrorism Act on September 9, 1998. Before this act passed in early 1999, President Clinton addressed the opening session of the fifty-third United Nations General Assembly in New York on September 21, 1998 to solicit international support to combat terrorism. President Clinton identified terrorism as the greatest threat to peace, not just to the United States, but also to the world. He stressed that terrorism is not fading away with the end of the twentieth century and that, "it is a continuing defiance of Article 3 of the Universal Declaration of Human Rights, which says, 'Everyone has the right to life, liberty, and security of person'."[62] This speech was one of many attempts to gain international support against terrorism. It may have been prescient as well.

Before the September 1998 speech to the United Nations General Assembly, a conference between the United States and the European Union (EU) was held on May 18, 1998. Its conclusions stated, "The United States and the EU member states are strategic allies in the global fight against terrorism . . . we oppose terrorism in all its forms, whatever the motivation of its perpetrators, oppose concessions to terrorists, and agree on the need to resist extortion threats. The United States and the EU condemn absolutely not only those who plan or commit terrorist acts, but also any who support, finance or harbor terrorists."[63] From this conference, all parties involved agreed that to end state-sponsored terrorism, international cooperation is necessary in the global economy.

Due to the success of the United States-EU summit, the State Department, as the lead federal agency designated to combat international terrorist threats, hosted an international counterterrorism conference on June 16–18, 1999. Each panel discussion included members from NGOs such as RAND and CSIS, and government officials including Israeli diplomats and leaders of NATO, as well as twenty-two United Nations member states. This conference focused on the need to cooperate across borders in order to eliminate foreign and domestic terrorists' threats. If any state openly sponsored terrorism, economically or physically, economic sanctions would be imposed from all member states of the EU and the United States. To do this, the Secretary of State maintains a list of countries that have "repeatedly provided support for acts of international terrorism," and these include Cuba, Libya, North Korea, Iran, Iraq, Sudan, and Syria. However, all of these efforts obviously pale before the current counterterrorism campaign being conducted around the world by the current White House. And as noted in the beginning of this chapter, the lines between offensive and defensive operations blur in the conduct of these missions. In fact, though there are obviously CTIO issues involved with Operation Enduring Freedom (OEF), the authors chose instead to put most of these new developments, specifically the war in Afghanistan under the offensive missions, later in the book.

PDD-63: *Critical Infrastructure Protection*
If *ER '97* and *JV 2020* are a two-generation change for IO events, then probably the key moment for the USG organizational architecture came with the promulgation of PDD-63. Although it has been touched on in earlier chapters of this book, we feel that it was of such importance as to merit a standalone section. This is because circumstances are forcing many military and government activities to develop defensive plans based on vulnerabilities rather than on the threat. The USG, specifically the Clinton administration, attempted to help that coordination by adding CIP as a new PDD, which was part of their overall NSS. Released on May 22, 1998, PDD-63, *Critical Infrastructure Protection,* this policy was also covered in the IO organization section as a critical new element in the Clinton administration's overall strategy. Much of the impetus for this new policy came from a series of well-publicized terrorist attacks such as Ruby Ridge, the World Trade Center bombing in 1993, Waco, and the federal building in Oklahoma City. All of these incidents and standoffs by federal agents led to a general feeling by the White House that there was a threat to our national infrastructures and, in turn, our

national security. The Clinton administration therefore believed that if they could emphasize the need for public and private cooperation for infrastructure protection, they could be successful from the current threats.

To that end, the administration began working in 1996 on developing a series of plans and policies that ultimately culminated in PDD-63. The most important document included in this evolution was Executive Order 13010 *President's Commission on Critical Infrastructure Protection* (PCCIP) and *Critical Foundations*. This report laid the foundation for *Critical Infrastructure Protection* (PDD-63), which is also tied very closely to another policy document, *Counter-Terrorism* (PDD-62). Both were released on the same day and were headed by the same director at the NSC, Richard Clarke. This physical and symbolic closeness was good in that it allowed the NSC to focus on all aspects of the issues for protecting U.S. infrastructure, as shown in Diagram 3-18. The initial goals of PDD-63 were to:

- Establish a national center to warn of and respond to attacks
- Address the cyber and physical infrastructure vulnerabilities of the federal government by requiring each department and agency to reduce its exposure to new threats
- Allow the federal government to serve as a model for the rest of the country for how infrastructure protection should be attained
- Seek voluntary participation of private industry to meet common goals for protecting our critical systems through public-private partnerships

Lead Agencies	Sector
Commerce	Information and Communications
Treasury	Banking and Finance
EPA	Water Supply
Transportation	Aviation, Highways, Mass Transit, Pipelines, Rail, Waterways
Justice/FBI	Emergency Law Enforcement Services
FEMA	Emergency Fire Service, Continuity of Government Services
HHS	Public Health Services
Energy	Electric Power, Oil, and Gas Production and Storage
Lead Agencies	**Special Functions**
Justice/FBI	Law Enforcement and Internal Security
CIA	Foreign Intelligence
State	Foreign Affairs
Defense	National Defense

Diagram 3-18   *CIP Lead Agencies by Sector*

To accomplish these goals, PDD-63 established the following institutions:

- The National Infrastructure Protection Center (NIPC) at the FBI which fuses representatives from FBI, DoD, United States Secret Service, DoE, DoT, the intelligence community, and the private sector in an unprecedented attempt at information sharing among agencies in collaboration with the private sector. The NIPC also provides the principal means of facilitating and coordinating the federal government's response to an incident, mitigating attacks, investigating threats, and monitoring reconstitution efforts.

- Information Sharing and Analysis Centers (ISACs) were encouraged to be set up by the private sector in cooperation with the federal government and were modeled on the Centers for Disease Control and Prevention.

- A National Infrastructure Assurance Council (NIAC) drawn from private sector leaders and state/local officials to provide guidance to the policy formulation of a national plan.

- The CIAO provides support to the national coordinator's work with government agencies and the private sector in developing a national plan. The office also helps to coordinate a national education and awareness program and legislative and public affairs.

In the original plan, two years after PDD-63 was released, an initial operating capability was supposed to have been achieved, as shown in Diagram 3-19. Within five years (2003), PDD-63 was to have the operating capability to fully protect the nation's critical infrastructures. This would include federal, state, and local sector compliance as well as commercial cooperation. As of this book's printing, PDD-63 appears to be on track. Numerous milestones have been passed, and with increased guidance from the NSC, *Critical Infrastructure Protection* has significantly changed the public-private cooperation over the last few years. Specifically, *Practices for Securing Critical Information Assets,* released in January 2000 by the Critical Infrastructure Assurance Office, and *Defending America's Cyberspace, National Plan for Information Systems Protection Version 1.0, An Invitation to a Dialogue,* released by the White House in 2000, were extremely important. Of course, it didn't hurt that there was a huge amount of attention paid to computer systems due to the Y2K rollover date, but much of the credit must be given to Richard Clarke and his staff. It was through their efforts that the success of PDD-63 over the last five years was assured.

Dept/Agency	1998	1999	2000
National Security	$975	$1,185	$1,403
Treasury	23	49	76
NASA	41	43	66
Transportation	20	25	51
Justice	26	54	46
NSF	19	21	27
Commerce	9	22	18
HHS	22	12	13
Other	9	18	37
Total	$1,144	$1,429	$1,737

Diagram 3-19 CIP Budgets by Department, Fiscal Year
1998–2000

Summary

In conclusion, information protection is a huge area that needs massive input from the U.S. government agencies and military to be effective. Officials have only recently—in the last few years—become aware of the true extent to which our nation is vulnerable to asymmetric attacks, and the attacks of September 11, 2001 only accentuated the vulnerabilities. As you have realized, a large number of programs have been covered in this chapter that illustrate how the various governmental agencies have attempted to "come to grips" with the real-time threats posed by terrorism with IO and now Homeland Security. Taken together, IA, CND, CTIO, and CIP are all key elements of a defensive IO program, and it is evident that a significant effort has been made to increase funding and importance of these areas over the last few years. If we can emphasize only one thing out of this whole chapter, it would have to be that the best defense against IO attacks must be based on people and training. No amount of technology and policy can overcome a determined foe, as was obviously shown at the World Trade Center and the Pentagon, but often times, with proper training, good managers and administrators can mitigate the effects of these attacks. And of course, only time will tell how effective these measures truly are.

CHAPTER 4

Information Projection

Shaping the Global Village

"Iraq lost the war before it even began. This was a war of intelligence, electronic warfare, command and control, and counterintelligence. Iraqi troops were blinded and deafened . . . modern war can be won by informatika and that is now vital for the U.S. and U.S.S.R."[1]

Soviet Lieutenant General S. Bogdanov

The last chapter discussed the need for commanders to protect their information and information systems; however, perhaps even more important is the need to plan military operations to exploit vulnerabilities in adversary information and information systems. JP 3-13 defines offensive information operations as:

> The integrated use of assigned and supporting capabilities and activities, mutually supported by intelligence, to affect adversary decision-makers to achieve or promote specific objectives. These capabilities and activities include, but are not limited to, operations security, military deception, psychological operations, electronic warfare, physical attack and/or destruction, and special information operations, and could include computer network attack.[2]

It is significant to note that this definition includes CNA, which before the publication of JP 3-13 in October 1998 was originally classified by the Department of Defense. In addition, significant discussions of offensive information operations are contained in JP 3-13 under the following phrase: "Offensive information operations can support defensive information operations."[3] Although this may seem a subtle change, it does support the further assimilation of both disciplines into one, reinforcing a more proactive, rather than reactive approach. It also correlates well to the American penchant for the idea that the best defense is a good offense. In addition, offensive IO has also concentrated on the two main

111

developments that have come to the forefront in the last several years. Involving CNA and perception management, both of these areas are where most of the focus for IO is currently being conducted.

Offensive Information Operations

The target for offensive information operations is the human decision-maker. This cannot be emphasized enough, as it shows the difference between the older disciplines incorporated in C2W. Whereas the earlier policy concentrated on nodes and links, IO has instead focused on influencing the commander or decision-maker. Commanders will plan to employ offensive IO capabilities and related activities with the goal of influencing their adversary's observation, orientation, and perceptions, thus causing them to decide to act in a way that is advantageous to that commander's military objectives. As mentioned in Chapter 1, within U.S. doctrine, this is called IS, and IO is a sub-component of that theory.[4]

For offensive IO, there are a number of planning considerations. First, commanders need a range of capabilities in order to shape their broad operational environment. For example, the former commander in chief of the U.S. Southern Command, General Charles Wilhelm, U.S. Marine Corps (USMC) recognized IO as a core competency and considers its employment as he would any other battlefield operating system when designing his theater security cooperation plan (TSCP), which is the new acronym for the TEP.[5] Because these plans are a tool for managing and shaping a CC's AOR for a period of seven years, these documents are especially useful in the IO context. Because so much of IO is done before a crisis occurs, and before a warning or an execution order is issued, it is the TSCP and the collateral Department of State reports that in essence constitute the IO attack plan. Therefore, IO can consist of offensive tools such as public affairs, civil affairs, psychological operations, and CNA—all of which can be conducted in a pre-hostilities phase of a potential conflict. In addition, offensive IO also considers both lethal and non-lethal weapons as means available to disrupt the adversary's information flow and services.

There are also some general principles for employing offensive information operations that must be considered in planning military operations. Offensive IO may be the main or supporting effort of a Joint Force Commanders (JFC) campaign or operation, but it must also support the overall military objectives and have some form of observable measures of effectiveness (MOE). Finding such measures is often one of the most problematic issues, simply because some aspects of IO do not lend themselves to easily quantifiable observations. As mentioned repeatedly, IO is not some-thing that can be done quickly or in a crisis mode. Therefore there is also a need for extensive lead-time in preparing for offensive IO in order to ensure that the adversary decision-maker is responding in the way one intends, thus raising the need for a thorough intelligence preparation of the battlespace (IPB) prior to any offensive IO effort. Also, to be successful, offensive IO must fit with overall U.S. security objectives and be consistent with established rules of engagement. Furthermore, offensive IO must also be thoroughly integrated with all those non-DoD organizations throughout the interagency involved with the particular operation. In a regional con-text, the CC and ambassador/country team relationship is the most crucial for organizing and conducting an offensive IO strategy. The need for long lead-times, plus the required interagency involvement, has thus to some extent inhibited the successful use of IO as a strategy to shape the envi-ronment. However, there are indicators that as IO becomes more estab-lished within the USG, these factors will become more accepted.

Therefore it appears evident that both doctrinally and operationally, the distinction between offensive and defensive information operations is becoming increasingly blurred. President Clinton, for example, communi-cated this sentiment in January 1999 in an address to the National Academy of Sciences. Discussing the implications of the threats posed by terrorists (including cyberterrorists) and weapons of mass destruction, President Clinton noted that, "because of the speed with which change is occurring in our society, in computing technology, and particularly in the biological sciences, we have got to do everything we can to make sure that we close the gap between offense and defense to nothing, if possible. That is the challenge here."[6]

By doctrine therefore, IO is composed of six capabilities and two related activities. We could discuss all of these mission areas in great detail, but it seems more beneficial to instead focus on the specific func-tions that are new or that have changed significantly in the last few years. Therefore, the first offensive IO capability that should be addressed is computer network attack or CNA, as shown in Diagram 4-1. This will be followed by space-based applications of IO, electronic warfare (EW), and international public information (IPI). These four areas will be the focus of this chapter because this is where we, as the staff of Information Warfare at the Joint Forces Staff College, think that the biggest advances in IO will come in the future. This is not to deni-grate PSYOPS, deception, OpSec, or physical destruction (PD), but instead to emphasize to the reader what we believe are the most impor-tant areas to emphasize.

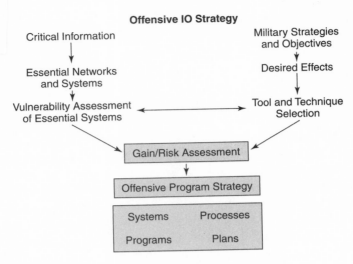

Offensive IO Strategy

Diagram 4-1 Offensive IO Strategy

Computer Network Attack

"Computer network attack." The very term evokes thoughts of cyberwar and futuristic technology, with visions of precision accuracy and war without needless violence, perhaps even a kinder and gentler form of warfare. Yet perception does not equate to reality. Therefore, though by definition CNA is a current warfighting capability of the United States, some would say that it is so limited by legal, political, and security constraints as to make it virtually useless to the combatant commanders.

Though this section discusses these issues and others associated with CNA, because much of this technology is classified, details about some of the subtopics are unavailable. In fact, the very term *computer network attack* was classified until October 1998 with the publication of JP 3-13. Before that time its use had been classified at least to the "Secret" level and even today, CNA cannot be associated with certain commands without immediately raising the classification level. For example, JP 3-13 gives only one sentence to CNA before referring the reader to the "Secret" supplement. So, unfortunately, for the present time, much of the discussion of CNA must be short and generic. This chapter will instead focus on what CNA can and cannot do and raise some issues about its capabilities and limitations. It does not, however, describe how the United States is conducting CNA missions, nor give examples of any recent CNA operations.

By definition CNA operations disrupt, deny, degrade, or destroy information resident in computers and computer networks, or the computers and networks themselves.[7] Most people think of CNA as being conducted over the World Wide Web or the Internet, but in fact the physical destruction of a computer system or network by kinetic means also qualifies as CNA. Indeed, if a computer is not connected to a local area network (LAN) or the Internet and is a stand-alone system, then it will have to be physically destroyed or have malicious computer code physically inserted into its software program. That form of CNA might no longer come under the purview of uniformed military forces, but instead might be designated to other organizations that have traditionally conducted such operations.

However, it is the ability to disable an enemy's computer system from afar, often from the safety of one's own command and control center, that makes this new form of warfare highly desirable. The safety and virtually risk-free concept of attacking from a distance has led some people to suggest that CNA is the "silver-bullet" that everyone wants in a new weapon. But, of course, there is no such thing as a "silver bullet." Thus, though CNA has enormous potential and capabilities, so far a variety of legal, political, and technological constraints have kept it from being fully utilized as envisioned.

In fact, the legal constraints alone may restrict the use of CNA by U.S. military forces. The caution is similar to the NBC (nuclear, biological, chemical) criterion of not using any weapon on an adversary that the user is not fully prepared to defend against. The United States is the most vulnerable of any country to a computer attack, so for us to initiate a CNA operation would surely open the floodgates against our nation as well. Likewise, CNA can affect civilian as well as military targets with the same equipment. Should an attack target only military bases or should it try to cripple the economic base of a nation? If an attack targets financial institutions and spreads panic among an adversary's populace, is the attacker operating within the Geneva Convention and Laws of War, specifically those that require attacking only military targets, while minimizing collateral damage and avoiding indiscriminate attacks?

The specter of CNA is so large that Russia has attempted to limit its use by the adoption of new international laws. A proposal has been submitted to the United Nations to ban the use of IO. Although the White House under the Clinton administration successfully deflected their proposal, it could gain strength and develop a life of its own—witness the worldwide movement to ban land mines. In addition to the legal constraints, there are also technological challenges in conducting CNA. To be effective, a CNA operation is aimed at a single computer or a system of computers that conducts a specific mission. The intelligence needed to conduct a computer network attack is an order of magnitude greater than what may be needed for a bombing mission. Some of the following questions need to be answered to prepare adequately for a CNA operation:

- Where is the system that is to be the target? What room on what floor of which building?

- What kind of hardware is hosting the system?

- What software is resident on the computer and which version is currently installed?

- Is the computer connected to the World Wide Web or is it "air-gapped," that is, a stand-alone system?

Once these and other questions are asked, the "painless" claim for CNA proves wrong, and these operations become as difficult or perhaps more difficult than other types of attacks that are more familiar. In addition, the targeting aspect of CNA is extremely difficult. Once intelligence has narrowed the list to a system or perhaps a single unit, the attacker must be sure that the targeted computer is in fact the right machine and not an intermediate Internet Service Provider (ISP). Thus, CNA is defined by the big four "D" words—disrupt, deny, degrade, and destroy, another facet of these types of missions is actually gaining access to a computer. That activity is often referred to as computer network exploitation (CNE) and, more often than not, this is the hardest part of CNA. Once in and having gained access, anyone can mess up a system, but getting past the security systems is definitely tricky. To confuse matters even more, though only certain organizations are allowed by law to conduct CNA, that same restriction does not necessarily apply to CNE.

So what are the signs that a CNA operation is occurring? Does a computer explode on the desk? Let's hope not. Does it simply stop working? Many experience the infamous "blue screen of death" everyday themselves, so is that a sign of a CNA attack, or is it just the usual operator or operating system error? A CNA operation is often much more subtle. The attack may occur without the adversary's realizing that it has been attacked. Because computers are often viewed as office rather than military equipment, they tend to not have the technical documentation or personnel support of a major weapon system. For example, suppose a computer system stops working. What does the operator do? Most will try to logically troubleshoot the computer by looking for obvious faults, and then, failing that test, try to reset the software. Perhaps they will hit the ESC key a couple of times or maybe try CTRL ALT DEL and end the particular task, or in the worst case, simply use the ON/OFF switch to reboot the system. This process probably takes about ten to fifteen minutes, and then the frustrated operator calls the system administrator, who, with luck, can respond to the trouble call in a timely manner. The administrator often goes through the process again, trying to reboot the system and checking its configuration. The whole process can take time, during which a vital message may have arrived; or an action may have taken place that has gone undetected because the computer was off-line. If IO is simply an integrating strategy that creates effects from different warfare areas, the denial of computer service is no different than jamming or destroying the same equipment. If one can safely affect that computer system from afar, and guarantee that one can control its effects, then perhaps CNA is the weapon of choice.

By law, the four uniformed services are required to recruit, train, equip, and support the armed military forces and provide those forces to the CCs for their use around the world. Therefore weapons procurement becomes the responsibility of the individual services, a major political and financial issue depending on the system. But suppose the CC directly acquires a "weapon" from outside sources, without service testing. The services may also have issues concerning control over software programs that have not gone through the typical acquisition process. Yet, if a CNA program is used by a CC, at some point it must be "weaponized." This can present major problems when services are not involved in the procurement or training of these CNA weapons and do not maintain them in any sort of inventory. Questions will arise about who owns them and who can use them. Other problems may arise, namely concerning the deconfliction of CNA operations. The services have developed the Joint Force Air Command and Control (JFACC) system to deconflict air missions among the different services, and it has worked reasonably well in a multitude of operations during the last decade. However, there is no comparable system for CNA operations. Instead, at the time of this book's publication, a number of classified groups currently meet to deconflict these types of computer operations.

The current perceived lack of use of CNA weapons could also be attributed to the fact that many senior officers are not familiar with them. They grew up in a military filled with kinetic solutions, and unless they are educated on the potential effects of these new weapons, their first decision is often to not use them. Likewise, if a CNA program is kept "behind the green door" in a compartmented cell and brought out only in a moment of crisis, its use will often not be approved. Senior leaders must be educated about programs that allow them to understand the capabilities of CNA. Only then can they appreciate its capabilities and be more inclined to use these weapons when the opportunity arises.

There are thus many complicated issues involved with CNA that make it hard to discuss at an unclassified level. Needless to say, this is a new and important warfare area of IO, one that will need time not only to evolve but also to prove itself to a whole new generation of military leaders for its use to become commonplace. The potential for the future use of CNA is great and with proper education, research, and development, we may be able to eventually realize that capability.

Space and Its Relationship with IO

Space has become an important factor in the arena of IO. What today is termed the "Information Age" is largely a result of the use of space as a catalyst in this very significant evolution. Many people today think of space as a far away place of the future, not realizing that satellites are passing overhead today at a mere distance of 100 miles and that the benefits of space are as close as their TV remotes, cellular phones, the nightly news, and a number of other daily conveniences. Space plays an integral role in all aspects of military operations as well as becoming an ever-increasing part of most USG departments. Thus the far reaching impact of space use has had significant results, changing the very world in which we live by providing an apparent shrinking and in some cases dissolving of international boundaries, compression of time, and the ability to have near instantaneous insight into happenings around the globe.

The benefits of space-derived information have played a significant role since the early days of the space program, and there has been a remarkable change in the last few years that promises to revolutionize how we look at space in the future. This change involves the greatly increased availability of space-derived data to the public and the implications that free access to space systems can have upon governments and their respective militaries. This distribution of power away from the military and official government agencies to NGOs and individuals has also considerably altered the access to this information. This idea ties directly to the theme discussed in the first chapter—the power of information is now no longer in the hands of the military and government alone, but has instead been distributed more to the people. This concept is truly revolutionary.

For the majority of time since the dawning of the space age, access to space-derived information has been limited to a very elite few. These early space systems provided an immense information advantage to leaders of the countries that owned these assets. The consumers of these systems benefited by being able to see into others' backyards, to eavesdrop on conversations, and to provide early warning of potential attack. The diplomatic and military value of this information is clear in a quote by Lyndon Johnson who is reported to have said, "We've spent thirty-five or forty billion dollars on the space program. And if nothing else had come out of it except the knowledge we've gained from space photography, it would be worth ten times what the whole program cost."[8]

Today, the availability of space system information has greatly increased with the advent of numerous commercial systems that have made space-derived data available to the public. This commercialization of space in recent years has brought astounding results as the application of space data finds its way into more and more aspects of daily life. Today, the commercial market provides high-resolution imagery, precision navigation, highly accurate timing signals, remote sensing data, telecommunications support, and a host of other applications. The commercial application of space information is so great that, indeed, the Information Age has come about largely as a direct result of capabilities provided by these systems. But in this case, the availability of information is a double-edged sword that is effectively whittling away at the advantage enjoyed by the United States as one of the historical few that has in the past controlled space system information.

One recent event in the commercialization of space systems involves space imagery. High-resolution imagery of less than three meters was not publicly available before the 1990s. With the September 1999 launch of the Ikonos imagery satellite, one-meter resolution imagery is now available to the public at a cost of $30–$300 per square mile. The impact of this increased availability is rapidly becoming apparent as new commercial applications of space imagery are identified. It also provides an excellent intelligence-gathering tool to any country that wants to have a better look at its neighbors or for terrorist groups to identify and monitor potential targets. Instead of government officials dictating the appropriate time to release information gleaned from a space surveillance satellite, the media, NGOs, or even normal citizens can now "beat them to the punch" by ordering and analyzing the appropriate images. There is even now the potential for government decisions and policies to be challenged by anyone armed with the appropriate surveillance information. Diagram 4-2 illustrates the recent advances in space imagery.

Space-based navigation is another example of recent changes in regard to the commercialization of space. The Global Positioning System (GPS) provides unprecedented accuracy and timing information as a free service, available to anyone with relatively inexpensive receiving equipment. Developed and deployed by the military in the 1980s, this system has obvious military and civilian applications. GPS was initially designed to provide

(a)

(b)

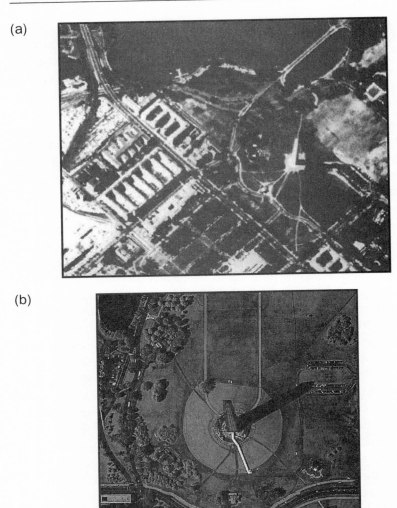

Diagram 4-2 Contrasting Images of the Washington Monument.
Picture (a) of the Washington Monument was taken on
September 27, 1967 by a military Corona satellite. Corona was
the nation's first photo reconnaissance satellite system,
operating from August 1960 until May 1972. Picture (b) on the
bottom also shows the Washington Monument and was taken
on September 30, 1999 by the commercial Ikonos satellite.

commercially available navigation, and the overwhelming number and breadth of commercial applications is unprecedented. Commercial GPS systems are beneficial for manufacturers tracking the status of deliveries, farmers improving land management applications, outdoor enthusiasts, police keeping tabs on dangerous criminals on parole, rental car options assisting customers to locate their destinations, schools providing advance notice to students as busses approach their respective bus stops, and even for fishermen finding their favorite fishing holes. The number of navigation applications continues to grow at a phenomenal pace, and arguably the greatest commercial benefit from GPS is derived from the highly accurate timing signal broadcast from each GPS satellite. This timing signal plays a critical part in telecommunications, electrical generation, and other technologies. Yet even with this extraordinary commercial demand, the most accurate GPS signals were historically reserved for strictly military use before the year 2000. On May 1, 2000, under new policy guidance from President Clinton, these highly accurate navigational signals were made available to all commercial users, resulting in an immediate ten-fold improvement for all commercial receivers.[9] Of issue was the availability of GPS navigation data, which under this proclamation now changed the advantage that our military forces had previously held with exclusive access to this system. The military applications of GPS data are very extensive, ranging from precision weapon delivery and force movement to the tracking of logistics and supplies. This advantage, which the DoD previously held, is now slipping as other countries realize the value of GPS and begin to acquire, field, and use ground receivers themselves. Though the U.S. military reserves the capability of selectively degrading the commercial GPS signal, it will never again enjoy the unchallenged availability to high-precision navigation that was available with the initial GPS deployment.

As commercial capabilities continue to grow, the United States needs to develop a strategy for handling the availability of space technology overseas while maintaining the national's domination in space. This issue becomes increasingly complex as new international consortiums enter the space arena and the number of commercial space-based sensors increases. This issue is further complicated as the capabilities of commercial satellites increase to the point that they rival or surpass national and military capabilities. This could lead to the loss of advantages enjoyed during previous conflicts, such as:

- The practice of deception will be complicated. The "left hook" tactic used during Desert Storm worked because the opposing side did not see it coming. With commercially available satellite imagery, this type of maneuver will become harder.

- The practice of hitting critical nodes to eliminate adversary C2 and lines of communication has now been greatly complicated. For the price of a low-end computer, anyone can purchase a handheld cell phone that uses ground- and/or space-based cellular technology. Utilization of this capability can provide a robust command, control, and communication infrastructure that is difficult to counter.

- Proliferation of the GPS has for all practical purposes eliminated advantages associated with space-based navigation. Integration of this technology into theater ballistic missile systems can greatly improve accuracy and place armed forces at greater risk.

Yet as recently as May 1999, the U.S. military, which possessed a tremendous intelligence and space-based photography capability, was still deceived by the camouflage tactics used by the Serbians. So to say that access to one-meter resolution pictures will negate the traditional military capabilities of the armed forces is a bit premature, we believe. However, the importance of space with relation to IO is emphasized by the shift in mission areas of CND and CNA to SpaceCom over the last few years. As mentioned earlier, UCP '99 directed this shift and the CC's staff have worked very hard during this period to operationalize and institutionalize both space and IO as warfighting disciplines. To the extent that they become effective and synchronized in the future will be interesting to witness. Of course the shift of SpaceCom to NorthCom and StratCom with the set-up of the Homeland Security Department in January 2003 ultimately changed that emphasis as well.

The Relationship between EW and IO

When JP 3-13 was published in 1998, it was generally thought that confusion over the controversial subject of IO would subside. However, as has been illustrated elsewhere in this book, that is certainly not the case. One area still causing problems is the relationship between EW and IO. In this chapter, the authors hope to clear away some of the confusion. We will do this by defining the architecture of EW and then use examples to show when EW is IO, and also when it is not. As mentioned in Chapter 1, the joint definition for IO is so broad that at times it can encompass nearly anything, but the attempt of this section is to give concrete examples of what IO is and also what it is not.

JP 3-13 defines IO as "Actions taken to affect adversary information and information systems while defending one's own information and information systems."[10] Given this definition, the question arises: "When we conduct electronic warfare, are we affecting adversary information and information systems or defending our own?" Perhaps the definition of EW will offer a clue. JP 3-51, *Joint Doctrine for Electronic Warfare,* defines electronic warfare as "Any military action involving the use of electromagnetic (EM) and directed energy to control the electromagnetic spectrum or to attack the enemy." The three major subdivisions are:

- Electronic attack (EA)
- Electronic protection (EP)
- Electronic warfare support (ES)

Which brings us to another question: Are the electromagnetic spectrum and the enemy considered information systems? Before we answer this question, we need to know what an information system is. JP 3-13 defines an information system as "The entire infrastructure, organization, personnel, and components that collect, process, store, transmit, display, disseminate, and act on information."[11] So for most people, this means that the electromagnetic spectrum is an infrastructure or conduit that facilitates the reception and transmission of information. JP 3-51 defines the electromagnetic spectrum as "The range of frequencies of electromagnetic radiation from zero to infinity." This definition alone certainly doesn't clarify anything; however, the doctrine further expounds on the concept of the EM environment. JP-51 further states that "today, electromagnetic (EM) devices are used by both civilian and military organizations for communications, navigation, sensing, information storage, and processing, as well as a variety of other purposes." The spectrum is thus an infrastructure over which EM devices transmit or receive information! So if we think about this logically, any military action to control the electromagnetic spectrum affects an infrastructure, which is part of an information system! Is that a radical change or what? In the first chapter we described the revolutionary change that IO was bringing to the state of warfare. This is proof.

To show the specific advantages of EW, the next few paragraphs delineate some current doctrinal questions that are arising concerning relationships between EW and IO. Electronic attack is defined in JP-51 as "that division of electronic warfare involving the use of electromagnetic energy, directed energy, or anti-radiation weapons to attack personnel, facilities, or equipment with the intent of degrading, neutralizing, or destroying enemy combat capability." A good example is if an anti-radiation missile destroys a radar antenna and the radar is an information system, this would be a case where EA is IO. If, on the other hand, we use a laser to destroy a ballistic missile, many would argue that this unguided missile is not an information system, and in this case EA is not an example of IO.

Moving on to electronic protection (EP), this sub-field of EW is defined in JP-51 as "that division of electronic warfare involving passive and active means taken to protect personnel, facilities, and equipment from any effects of friendly or enemy employment of electronic warfare that degrade, neutralize, or destroy friendly combat capability." If, for example, the personnel, facilities, and equipment are part of an information system, then EP is IO. Let's say we reprogram our radar warning receivers to recognize an enemy wartime reserve frequency. In this case, we've protected our information system and therefore, EP is IO.

Electronic warfare support (ES) is defined in JP-51 as "that division of electronic warfare involving actions tasked by, or under direct control of, an operational commander to search for, intercept, identify, and locate or localize sources of intentional and unintentional radiated electromagnetic energy for the purpose of immediate threat recognition, targeting, planning and conduct of future operations." ES is not normally considered a part of IO because collecting, processing, and disseminating information falls under one of the three components of information superiority called "relevant information." The other two components are information systems and information operations, as was discussed in Chapter 1.

Earlier, we made the case that EW and IO overlap. We started out by looking at the definition of IO and EW to see if we could clarify the relationship. Once we got beyond the overall definitions and delved into the subdivisions, we could clearly see that there are cases when EW is IO and when EW is not. The authors believe that this whole discussion is important because it shows how a particular capability relates to IO. It also shows that though there are eight capabilities and related activities. Just because a certain capability is conducted, in this case EW, does not necessarily mean that IO is being used as well.

Perception Management

Why in this new era, in this information age, do we not have a single point or executive agency to coordinate and control the flow of information within the USG? It is most likely that because information is a tool that organizations use and because its control is no longer singularly in the hands of the government, it is understandable that no one single agency or unit owns the sources or means of delivery, nor the production as well.

These changes in technology alter, and in some sense diminish, the importance of the traditional role of the executive departments. Likewise,

NGOs and other players have become more important in the last ten years because of their ability to move with speed and agility to the different crisis areas of the world. Therefore, in this post-Cold War era, the fungibility of information and the use of "soft power" by the United States has greatly increased the need for a mechanism to coordinate a coherent message.[12] Yet at the same time, the drastic downsizing of the DoD and State Department agencies that coordinate these activities has forced the need for a reorganization and reconstitution of USG public diplomacy capabilities. The unfortunate result of all this is that at this time, no single agency is empowered to coordinate U.S. efforts to sell its policies and counteract negative publicity. Therefore, the promulgation of PDD-68 *International Public Information* was an outgrowth of efforts by the Clinton administration to tackle the public relations portion within the greater issue of managing complex contingencies. However with the change in administrations in January 2001, this new policy guidance was officially rescinded, and never really lived up to the expectations nor the predicted positive response for which it was signed.

History of IPI

The history of PDD-68 and IPI within the Clinton administration is relatively short. During Operation Desert Storm, an information coordination cell (ICC) was set up at the Pentagon. The NSC played a major role in this unit and many people within that organization clearly recognized how important this ICC was and that it should be a continuous feature, not an ad hoc creation. Therefore in 1993, the DoD requested that the NSC create an ongoing information-coordinating body. Although nothing came of this initial effort, personnel under the Clinton administration, who were involved in this initial effort, and were still a part of the civilian infrastructure at the DoD later on, helped out with the actual creation of the IPI document.

As the Clinton administration became heavily involved in complex contingencies like Haiti and Rwanda, questions began to arise within the White House and the NSC concerning information's ability to intervene in order to change the power structure in a crisis situation. Basic ideas such as "all crises start with information" or "one group is controlling all of the information" began to be seen as possible methods for exploitation by the United States. Following these complex contingencies, a new policy document was issued entitled PDD-56 *Managing Complex Contingency Operations*.[13] The idea behind this new guidance was to build an executive committee (ExComm) that could coordinate interagency issues concerning complex contingencies around the world. It was realized that these problems could not be solved by military, economic or diplomatic power alone and thus other essential agencies needed to be brought into the decision process.

Concurrently with this development in mid-1997, a number of mid-grade White House and NSC staffers began to write a PDD on information as a public diplomacy tool. They wanted to use the PDD-56 construct of complex contingencies to build a permanent working group at the deputies level that could meet on a regular basis to formulate public information policy. Their goal was to assess the information projection capabilities of the different government agencies and evaluate their usefulness. By the spring of 1998, the IPI group completed its five-month assessment and concluded that a central coordinating body was necessary to integrate various federal departments' information-projection capabilities. The IPI working group then drafted a rough PDD document and routed it through the various governmental agencies for review. Throughout the rest of 1998, the IPI group was busy working on information issues within the interagency process. Due to budget cutbacks and a restructuring of the State Department, it was announced in late 1998 that by October 1, 1999, the premier public diplomacy organization in the government, the USIA, was to be merged with the DoS. Thus, concurrent with the review of the IPI PDD came a need to build a home for this group within the restructured State Department. Therefore, a key component of the State Department Reform and Restructuring Act of 1998 was to form a new under secretary of state for public diplomacy and public affairs. This new office was proposed to be the key component and champion for IPI and would chair the interagency core group once the policy document was signed.

Outside Influences on IPI

Thus on April 30, 1999, PDD-68 was signed by President Clinton, but without the usual fanfare. It was not detailed in a major policy speech or statement and in fact to this day, you cannot download a copy of the whole document off the World Wide Web like some of the other PDDs. The document is not considered classified but has remained for official use only and is a decidedly low-key policy statement compared to other PDDs issued by the Clinton administration. There are a number of reasons for this. First, IPI is a controversial issue. Anytime that you talk about trying to conduct psychological operations or perception management issues on a foreign audience, you are bound to attract controversy. Second, IPI became a victim of bureaucratic change within the State Department.

Because PDD-68 was handed over to a cabinet agency to coordinate instead of kept at the NSC level, it was influenced by the department politics more than perhaps other Clinton-era IO policy documents have been. The fact that the IPI core group was assigned to State, which is the main diplomatic agency for the USG, it is not, however, an operational command and that has somewhat prevented IPI from shaping the environment using information as was originally intended. A good example of this was seen in the actions of the IPI core group. The IPI core group was supposed to be the main interagency body that coordinates information issues, yet the group met infrequently and was never the tool that was originally envisioned. A number of reasons exist for this, but mostly it was due to the fact that key IPI visionaries were not able to remain concentrated on this issue due to circumstances including bureaucratic and organizational politics.

In addition, outside influences also affected the operation of PDD-68 in the last years of the Clinton administration. These were evidenced mainly in two articles that were written about IPI in August 1999 and published in the *Washington Times* newspaper. Both of these articles criticized IPI as simply a "smoke screen" by the White House to try to conduct psychological operations against the American people. It also did not help that PDD-68 was considered to be somewhat sensitive; therefore it was not generally available to the American public, so a much-needed general debate did not subsequently occur. These negative stories were released concurrent with the appointment of a new under secretary; so altogether, they tended to cast a pall over the whole issue. In addition, a final issue that delayed the implementation of IPI was the slowness of the reorganization at the Department of State, which prevented an essential piece of PDD-68 implementation from occurring. Namely, according to the policy document, a DoD liaison officer should have been assigned to DoS to work solely on IPI issues, yet that billet was gapped for the first fifteen months of IPI's existence.

Therefore, as you can see, the history of PDD-68 is relatively short. Much of the interest and heavy involvement from interagency officials all occurred over a three-year period. Today IPI is, for all intents and purposes, dead. Though the current Bush administration has retained the under secretary of state position, the initiatives begun towards the end of the Clinton White House have not been followed up. One cannot really speculate on the impact of IPI and the use of information as tool for preventing complex contingencies, but it will be interesting to see how any follow-up IPI policy fares in the future.

What Is IPI?

Simply stated, IPI includes a combination of public affairs, international military information, and public diplomacy. In reality, it is about the power of information, which is what this book is all about.[14] Public diplomacy is the open exchange of ideas and information. It is an inherent characteristic of democratic societies and is central to U.S. foreign policy initiatives, and as such remains indispensable to achieving our national interests, ideals, and leadership role in the world. Since the end of World War II, the United States has attempted to use public diplomacy as a tool to influence foreign audiences around the world. Thus, the development of PDD-68 is the latest in a long line of attempts by the USG to harness the power inherent in information.

Public affairs and international military information (IMI) comprise the other parts of IPI. Traditionally these two disciplines have not normally been associated with each other. Public affairs are often considered a support function. If you look at doctrine, public affairs is portrayed as a coordinating skill or relevant activity. It is not often seen as an offensive weapon or enabler that the military can use. IMI on the other hand, is a useful acronym for psychological operations. Also known as PSYOPS or perception management, this warfare area has a long and distinguished military history. It has successfully been used by armies throughout the world to create conditions mutually advantageous to combat operations. It has also been discussed previously to describe the importance of counterterrorism operations.

A highly sought after capability by military forces, PSYOPS is not a property normally associated with the State Department and the USG, other than the DoD. In addition, public affairs officers are often loath to associate themselves with PSYOPS in order not to "taint" themselves. This is because public affairs officers are always supposed to tell the truth and they should not be seen as trying to manipulate any audiences. PSYOPS, on the other hand, is all about trying to manage the perception of people, especially the adversaries' minds. As demonstrated in Operation Desert Storm, a properly conducted PSYOPS campaign can be a huge force multiplier—during that campaign, leaflets and radio broadcasts induced thousands of Iraqi soldiers to surrender to coalition forces. Thus, the use of information to manipulate the adversary's mind goes directly against many of the principles inherent in the public affairs profession. Yet the two disciplines have many similarities, and if coordinated correctly there are huge gains to be made by an organization that integrates information across the board as a part of an overall campaign.

What Was the Clinton Administration Attempting to Do with IPI?

IPI was the Clinton administration's attempt to not only combat negative propaganda by other nation-states but to also show in a truthful and united front what the policies of the United States were. As doctrine, it was designed to:

- Prevent and mitigate foreign crises around the world
- Collect and analyze foreign public opinion on issues vital to the United States
- Enhance the use of information assets

As mentioned earlier, public diplomacy is not new. An important factor of U.S. policy in the Cold War, information was disseminated throughout this period to foreign audiences by television and radio broadcasts, in a form of state-to-state dialogue. And we were not alone. Nations throughout history and to this present day have tried to use information to influence other countries as well as their own citizens. How successful they were in those attempts often depended on a number of factors, including cultural and psychological biases as well as the means and methods of technology used to transmit that information.

Therefore, as envisioned by its writers, PDD-68 was important because it was an attempt to develop another tool that could be used in shaping and preventing complex contingencies. As outlined in the earlier policy of PDD-56, the Clinton administration recognized that it needed to do a better job of promulgating the truths about the United States to the world. Though public diplomacy had worked relatively well in the Cold War era by using USIA officers and policies, the environment changed drastically in the last decade. Without the overwhelming threat from the Soviet Union, that agency lost much of its *raison d'être* and thus by the late 1990s, there was a move to absorb the independent agency into the State Department. Likewise, the incredible advances in technology and the explosive growth of NGOs have drastically changed the methods of state-to-state contact over the last few years. Therefore, methods that had been effective for conducting public diplomacy during the Cold War are no longer considered viable in this new era.

With the promulgation of PDD-56, the Clinton administration was attempting to develop an organization that could form in a crisis and help coordinate across the interagency and coalition boundaries to help prevent more debacles like the massacre in Rwanda. There was a belief that an

ExComm would develop the trust and mutual understanding among the key principals and deputies of the executive branch. Communication and compatibility are essential ingredients, and the ability to promulgate that information to the world is another one as well. If the United States could prevent contingencies by using information to shape the environment, then that would help the USG in many ways. PDD-56 was consistent with the NSS and DoD doctrine, specifically the national military strategy (NMS) and IO, as outlined in JP 3-13. These documents stressed that with the downsizing of the U.S. military forces in the last decade, it was imperative that attempts be made to minimize the need to deploy them around the world. If crises or contingencies can be contained, minimized, or perhaps even avoided through the skillful use of information, then that is usually the preferred option. However, PDD-56 is limited in the fact that the ExComm is only created at the start of a contingency. What is really needed is a standing organization, one that could use information in the pre-crisis phase to shape the environment and perhaps prevent a crisis from ever occurring. This is the rationale and grand theme for PDD-68, as shown in Diagram 4-3.

Therefore IPI was an outgrowth of the earlier doctrine and is, in fact, a required follow-up piece to the *Managing Complex Contingency Operations Policy.* It was written to fill that void for dealing with foreign audiences that was addressed by PDD-56. The objective was to improve the ability to prevent and mitigate foreign crises while at the same time promoting understanding and support for U.S. foreign policy initiatives around the world. Specifically, the Clinton administration wanted to avoid mistakes like those in Bosnia and Rwanda that may have been prevented if an effective information policy had been in effect. A device or

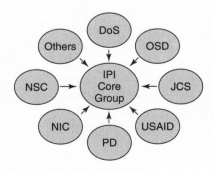

Diagram 4-3   IPI Key Organizations

group was needed to address the misinformation and ethnic incitement that had characterized these two regions. This group had to be interagency in character, in order to maximize its ability to develop a sound program, and it is called the interagency IPI core group (ICG).

PDD-68 was not a stand-alone product but instead part of a larger product. The broader theme for IPI was stated PDD-56 *Managing Complex Contingency Operations*. There were a lot of components involved with this process and IPI was only one of them, yet the idea was to form a group to attempt to integrate information as part of the interagency process. The founders of IPI wanted to make IPI a standard way of doing business, a regular part of the political military plan, not an "extra" that was added on at the last minute. This last point is key, because basically that is what the authors have been "preaching" from the beginning. Namely, that IO is not an added on military function, but instead that it is a truly revolutionary way of conducting operations and also, that it must be an integral part of any campaign, from the very beginning. Only then will the true advantages of information warfare be realized.

Why Has IPI Been "Less Than Successful"?

Why have the State Department and other interagency organizations been slow to adapt the IPI structure? Mainly because there are two important sensitivities connected with PDD-68. First, because the ICG has a connection with the intelligence community through the National Intelligence Coordinating Committee, this information group can draw on foreign intelligence sources. That means that they were, at that time, one of the few activities that had access to all-source intelligence. Therefore the ICG was in a very unique position, although other units like the State Department Intelligence Bureau also attempts to conduct all-source intelligence analyses through open-source material. The second sensitivity involves the interface between public diplomacy and public affairs. The professionals that conduct public affairs are very wary of both IPI and PD in general, and the PDD-68 initiative in particular. There was concern that the PA community could be tainted or affected by a close association with these other warfare areas. This line of reasoning was explained in two *Washington Times* articles that theorized that PDD-68 was a "Trojan horse vehicle" to PSYOPS the American public favorably to various Clinton administration foreign policy initiatives.

These are all noble ideas. Nevertheless, it still does not detract from the concept that IPI involves PSYOPS elements, which concerns many people. As reported in the *Washington Times* article "Professor Albright Goes Live" by Helle Bering (8/4/99), the State Department has not given up on educating the American public. The author indicated that she believed that PDD-68 was just the latest attempt by the Clinton administration to educate or "persuade" the citizens of the United States. The stated purpose of the IPI system was to coordinate all PA output from the different USG agencies. Clearly that is an unrealistic goal and far from doable, but because it was reported in the *Washington Times,* it gained credibility.

During the Cold War, most American citizens understood who the enemy was and why we were conducting military operations or foreign policy initiates. When the USG strayed from the course of containment or belabored the point too long as in Vietnam, public opinion would rise up to influence the politicians. However, the post-Cold War era is different. The United States has no peer competitor and therefore much of the American populace does not understand why the government is involved in these small nations in Africa or the Balkans. Therefore, the State Department often feels that it must demonstrate why this particular cause is important or is in need of attention, with so many others to chose from. Moreover, it is often a hard sell.

The State Department can only do so much educating of the American people. By law, the Smith-Mundt Act of 1948 prohibits the USG from targeting Americans with information that is aimed at foreign audiences; that is, you cannot conduct public diplomacy on your own people. That may have been relatively easy to separate via different media channels fifty years ago, but today with the merging of the telecommunications, computers, and media technologies it is much harder to ensure that the American populace will not consume information aimed at foreign audiences.

Detractors of IPI did not believe that the United States government was trying to abide by the Smith-Mundt Act and stated that PDD-68 was merely a tool to propagandize the American public. What this and the other critical article in the *Washington Times* failed to note was that the Clinton administration, though made up of some political appointees, was mainly supported by a vast bureaucracy of civil servants and professional government officials who did not owe their allegiance to any particular party. They are the backbone of the USG and for good or bad, they will survive any political administration. The civilian bureaucracy is divided into many different agencies and activities and, as is typical throughout the world, these departments all compete for budgeting dollars with each other. Therefore, in a sense, within the interagency process it is extremely difficult to reach consensus, much less to produce this Orwellian conspiracy against the American people. The career civil service personnel are often influenced more by internal department politics than the larger domestic agenda and therefore not only will it be hard to manipulate them, they also will be acting within an organizational and bureaucratic context.

For an IPI campaign to succeed, you therefore cannot wait until hostilities have begun. Instead you must begin earlier to mold and shape the political environment. Unfortunately in the case of PDD-68, that probably meant that IPI was written to the lowest common denominator. The policy document that ended up being published was very vague and contained no lashup, or mechanism, to coordinate with the Department of Defense. PDD-68 was essentially a concept without a structure, which probably doomed it from the start. Although it was supposed to be a component of PDD-56, that policy document was only a process to be used in contingencies. What was really needed for IPI to work was a coordinating group that could meet all the time.[15]

However, the most important thing that PDD-68 did was that it created an interagency dialogue during peacetime. Although this may seem a relatively easy task, it is in fact not at all easy. By trying to force the government to be proactive instead of reactive, IPI may have had more success than many people realize. It was forcing staffs to try to integrate information into the TSCP, as well as to work on horizontal integration issues across the board. Who knows, maybe if given a chance, and if Gore had won the presidential election of 2000, then PDD-68 could have done what it was designed to do from the beginning, namely, to look into the future at potential hotspots and use information as a shaping tool.[16]

Current Bush Administration Efforts

Even before the events of September 11, 2001, there had been efforts by the White House to update and rewrite a new NSPD to focus on perception management at the strategic level. A Defense Science Board on Managing Information Dissemination (DSB-MID) was published in early 2001, which had been developed by public diplomacy professionals and led by its Chairman Vince Vitto during the transition period between the Clinton and Bush administrations. However, the terrorist attacks and Operation Enduring Freedom (OEF) have pushed those efforts into high gear, as the effort of selling the American message has literally become a matter of life and death. Thus, much of the current changes in policy are directed as part of the global war on terrorism being led by the Bush administration around the world.

After the attacks of September 11, 2001, the State Department was looking to the National Security Council (NSC) for guidance on a strategic information campaign. Unfortunately, however, leadership was slow in forming. At that time, the Clinton-era NSC document, PDD-68 *International Public Information,* had been effectively muted, so there was no office dedicated at NSC to conduct a strategic perception management effort. The White House ended up, during major portions of this crucial period, simply contracting out their perception management campaign to The Rendon Group, a civilian company that specializes in strategic communications. John Rendon and his organization have been involved in many operations around the world since the first Bush administration.[17]

Gradually, as the campaign on terrorism continued throughout the winter of 2001, a number of influence plans and strategies were developed, yet the nascent NSPD still remained in a holding pattern within the interagency process. Eventually the decision was made to attempt to set up a Policy Coordinating Committee (PCC) based on the NSPD-1 type memo. The NSC then set up a strategic communications staff to coordinate this PCC within the counterterrorism (CT) framework. A process was developed to conduct strategic communications operations around the world, using a synchronization matrix similar to the IO cells in all the different CCs. In addition, an important task of this PCC is to conduct education and training of senior leaders, so that they can understand the power of information as it relates to CT. *Axis of Evil, Infinite Justice,* and *Crusade* are great examples of Bush administration public diplomacy missteps, where these actions and words have seriously hurt the White House in its global war on terrorism.

In the meantime, during the immediate aftermath of the terrorist attacks, Alistair Campbell, the communications director for Tony Blair (UK) had suggested to Karen Hughes, communications director for the Bush administration, to form a series of communication information centers (CICs) to concentrate on getting the pro-American message to the world media. Eventually three were set up, one each in Washington, London, and Islamabad, and in fact, the facility in Pakistan was actually an old USIA building. These groups perform admirably, focusing on public affairs and public diplomacy, working with domestic and foreign presses on their time cycles during the early phases of OEF. In fact, as alluded to earlier, John Rendon and his associates had been pushing all along for a faster response by the White House to the media. He believed that you should attempt to win every news cycle and not let a report go unanswered.[18] So eventually the response by the USG got better, but it should have been recognized much earlier. Foreign media needed to be addressed, and the fact that it took so long to get key USG personnel available decreased the efficiency and positive impacts possible.

Yet after OEF, these CICs were disbanded. However, before she left the Bush administration in late 2002, Karen Hughes started the process of setting up the Office of Global Communications (OGC) to force the public diplomacy community resident within the DoS and in the field to do a better job of explaining overall U.S. policies. Originally called the Office of Global Diplomacy, it was proposed out of frustration with the efforts of *Foggy Bottom*. This office will work in conjunction with the strategic communications PCC, as well as the DoS, to better coordinate strategic perception management campaigns.[19]

However, there was still thought within the DoD that more needed to be done to counter negative American publicity in the media. In February 2002, it was announced that the Office of Strategic Influence (OSI) was also set up by DoD, in an effort to coordinate its strategic perception management campaign. Comprised mostly of U.S. Army PSYOPS and U.S. Air Force special operations (SpecOps) personnel, plus a few DoD civilians, their mission was to respond to and counter hostile propaganda, using mostly human factors and a little technology.[20] At a meeting on February 16, 2002, Secretary of Defense Donald Rumsfeld approved the unit; however, the senior DoD public affairs official Victoria Clarke non-concurred. On February 19, 2002, the first article critical to the new organization was released while both the Secretary and Clarke were in Utah at the Winter Olympics. It was reported that Rumsfeld was livid but could not do much due to the political concerns. As satirically reported by Mark Rodriguez in the *Washington Post* electronic journal *Insight,* the demise of this DoD office was a political turf-battle with Clarke leading a disinformation campaign to retain control of all PA efforts.[21] Politically embarrassing to Secretary of Defense and the President, it was very comical to watch the government officials deny that the United States conducted strategic perception management campaigns. Every nation participates in these activities, but almost all deny their existence. Even foreign news agencies put a satirical touch on their reporting as they watched the American officials attempt to explain away the obvious.[22]

All of these organizational shifts allude to a question that has arisen recently, over the last five years. Where exactly should a strategic perception management campaign office be located? PDD-68 put international public information (IPI) activities at State, where they foundered for two years due to lack of budgetary authority and manning. In addition, the interagency process also hampered the ICG. At one time, they were sitting in on up to six different meetings, but progress was very slow. Originally the new draft NSPD recommended setting up the strategic perception management capability at an office in the NSC, but the White House has deleted the language for a separate NSC office of strategic communications and information. In addition, the older DSB-MIB had reiterated their desire to keep the perception management authority at DoS. The argument for keeping the PCC at NSC was basically because that organization is in a steady state. It looks at international affairs and foreign audiences in an operational manner, which was greatly missing from the IPI way of doing business. So there is strong logic for this. On the other hand, the argument for putting the PCC in DoS was led by David Abshire, who believed that a Tom Ridge-like figure was needed to drive this program.[23] In fact, during the Bush administration, Karen Hughes moved from her role as information director to counselor to the president, however with her departure from the White House, these efforts may have lost their momentum.

Therefore, the Office of Strategic Influence was set up because of this perceived leadership void, with ASD/SOLIC in the lead. It seemed to work well because it had money and also, because it was DoD organization, it didn't have to get DoS approval.[24] OSI had been placed at DoD, not at IPI because some believed that it was easier and more operational, which may have been the ultimate rejection of PDD-68—the overall belief that the strategic perception management campaign had been put in the wrong place by the Clinton administration, and that instead, an office should have gone to DoD or NSC instead.

The debacle concerning the OSI in February 2002, has, in the end, hamstrung a lot of the Bush administration's attempts to develop a strategic communications effort, and essentially this controversy has put the NSC PCC on hold for much of its tenure. As mentioned earlier, the term *strategic influence* was changed to *strategic communication* and the PCC was put under the CT umbrella at the NSC.

Thus the current structure of the new PCC along with the fledging OGC constitutes an attempt by the Bush administration to develop a public diplomacy potential. This is very ironic, because it was less than five years ago that the USIA was dismantled and its functions shifted under the greater umbrella of the DoS. In fact, in March 2002, Representative Henry Hyde (R-NY) proposed the reconstitution of that agency in his legislation *Information Protection Act of 2002*—HR 3969, to bring back capabilities which had so recently gone away. Much of this legislative proposal mirrors efforts by the DSB-MIB working group, however, to date, the State Department has not agreed with this concept, and in the long run, nothing may come of it in the interagency turf battles.[25] That doesn't mean that Congress is stopping its efforts, though, because bills have been forwarded in 2003 as well. In the end, the demise of the USIA may have contributed more to the failing of PDD-68, and thus the need for a new office in NSC, a new NSPD and PCC, more than any other action to date.

For in the end, it is not a new organization that will drive a strategic communications effort but instead a shift in the mindset of the White House and the NSC. The need to push senior officials to conduct briefings at 7 AM EST, to match Middle Eastern news cycles, or to ensure that U.S. Arabic speakers are available on Al Jeezra TV stations, is becoming a much more logical method of doing business. These ideas are now conventional wisdom as the value of strategic communications rise within the Bush administration. To be effective, one cannot just think in news cycles (24/7 around the world), but instead also in decades, so that the USG can be much more effective in a strategic management campaign.[26] In effect, there needs to be an issues agenda versus a value agenda. We need to take a short- and long-term approach to these problems, but it must also be led from top-down, with full White House and NSC leadership to ensure full interagency participation. It is only then that a true strategic perception management campaign will succeed, and the power of information will be realized.

IO in Operation Enduring Freedom

Compared with his father, George W. Bush had a much tougher road before him in developing and holding together a coalition for Operation Enduring Freedom (OEF). Unlike in Desert Storm when thirty-eight nations came together to liberate Kuwait, the new coalition that coalesced around the threat of terrorism is vastly different. For one thing, there is no one common enemy. The terrorists that comprise the al Qaeda organization are just one part of the overall equation. Based around the world, including within the United States, as well as many other "friendly" nations, in OEF there is no country or nation-state to defeat. This leads to the second fact, that the frontlines for this new "war" are everywhere, and nowhere, so there are few armies to command or forces to contribute. Third, this new coalition is huge, numbering almost 200 nations, which, to one extent or another, volunteered to support the United States in its cause. Fourth and finally, because the operation is so complex and varied, in essence not every nation is being asked to contribute in the same manner. Some are providing military assistance, others intelligence information, still others are tracking the financial links. In essence, there are four coalitions operating together, each with a distinctive but overlapping mission.

Taken together, these differences are already shaping a different picture of how these alliances are going to look and operate as compared to the coalition in Operation Desert Storm. In fact the overall military portion of the war on terrorism may only amount to 15–20 percent of the activities, with a much broader amount of attention directed at conducting legal investigations or freezing the terrorists' financial assets.

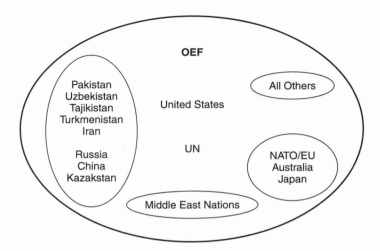

Diagram 4-4   OEF Venn Diagram

Think of it as a Venn diagram, with four circles or groups inside the larger overall structure of the coalition (see Diagram 4-4). Each of the different alliances are crucial in their own way, and in the end, success may not be measured by whether Osama bin Laden is killed or not, but instead by the ability of the Bush administration to hold it together for the duration of the campaign. Because it is so large and diverse, with a huge scope, there will be huge pressures on some nations to defect. These members are vulnerable, and if key states are lost, it could perhaps lead to eventual disintegration of the coalition. A form of prisoner's dilemma if you will, the sheer amount of "Realpolitik" currently being practiced by the administration may ultimately be its saving grace.

This sensitivity displayed by an earlier Bush administration will be needed again for the current "war." This new conflict is unlike any that the world has ever witnessed before, with transnational actors playing center stage. This is not going to be an attack on a single nation, instead it will be a fight on many fronts, including in our own backyard. There are a number of nations that may not want to recognize the threats to their regimes or governments, and so in this conflict, these leaders will be asked to make tough choices, decisions that may alienate them from much of their constituencies. But the final question that must be asked is simply: If 3,000 innocent people are not worth dying for, then what is? So it is a fine line that the Bush administration must walk, between demanding allegiance for the coalition and allowing nations to decide on their own in what capacity they will provide services and assets for the good of the alliance.

The Structure of the OEF Coalition

The strength of the alliance for Operation Enduring Freedom lies in its structure. As mentioned earlier, it is essentially four different coalitions operating together. The first circle is comprised of those states that border on Afghanistan itself: China, Pakistan, Russia, Uzbekistan, Tajikistan, Turkmenistan, Kazakstan, and Iran. All of these nations are invaluable for providing logistic and intelligence support for the United States, but the extent to which they do so will differ significantly. In each case, there will be a price to pay, and it will be up to both the Bush administration, as well as the leaders of these respective nations, to see if they can afford the costs. For example, Vladimir Putin from Russia played his cards very well, not giving away much and extracting concessions on a wide range of issues including, most significantly, his military operations in Chechnya. In addition, there is the possibility that he may also use his new-found power to leverage concessions from the United States in the form of limitations on deployment of a national missile defense system. General Pervez Musharraf of Pakistan, however, is in a much more difficult situation. He clearly did not want to not be seen as aligning with a terrorist group, as well as receiving the added benefits of aid from the United States, but his government may not be strong enough to force these opinions upon his public.

Though Pakistan is probably the most crucial nation in this first group, it may also be the most fragile. Access to this nation's airspace is crucial for sea-based strike assets as well as those operating from Diego Garcia. Pakistan is one of only three nations that formally recognized the Taliban government, and they hosted an ambassador from the fundamentalist state in Islamabad until late November 2001. Many clergy in the nation agree with Osama bin Laden, and religious schools dot the countryside, full of future soldiers for the Taliban army. Though Musharraf sided with the United States in the war on terrorism, there is no evidence that his constituency will follow.

The second ring of nations in the alliance consists of states that have the closest ties to the United States. These include members of NATO, the European Union, Australia, and Japan. All of these countries pledged military and/or economic support, but perhaps more importantly, for the first time, they all worked together to share criminal data on the terrorist organization al Qaeda. These countries are the willing, the capable, and the able. It is generally agreed that fewer nations is better. It is also understood that this is not a war that can be planned, so in essence this military alliance must remain small and flexible. To date, these like-minded states have developed a community of interests that goes beyond the normal bilateral relationships between the nations. Because much of these countries' financial and communication links are actually in Europe vice the Middle East, it is this unprecedented cooperation between these nations in the criminal and legal arenas that may have the most significance for the continuing capture of suspected terrorists.

The third ring consists of the Islamic and Arab nations who must come to grips about where they stand within the coalition structure. One potential problem for this is that in general, there is little "community of interests" on numerous foreign policy issues between the United States and Arab nations. Some scholars cite the lack of common values and little understanding of each other. Often times there is no filtering between nations on critical issues and everything is either black or white. For instance, President Bush stated on September 20, 2001, in his joint address to Congress, that this fight against terrorism was worldwide and that you are either with the United States or against it. That put a lot of pressure on these Arab nations because many of these are similar to Saudi Arabia and Pakistan, which have large internal dissident groups that are critical of the ruling leadership and regimes. These are precarious governments that are fragile and are vulnerable to domestic backlash. These nations are perhaps the most pivotal and crucial to the success of the coalition, for the very fact that they are Muslim nations and are vulnerable themselves.

The fourth circle of allies consists of all of the other nations that will play hugely disparate roles in the war on terrorism. Some of these nations provide limited military support, while others are conducting financial and criminal investigations. Some have legitimate concerns, while others do not want to be seen as unwittingly aiding the terrorist groups. For most of these nations, commitment is in the form of rhetoric alone, so they are in effect bandwagoning on the U.S.-led coalition. As George W. Downs argues in *Collective Security Beyond the Cold War,* coalitions are often only effective when a strong nation is willing to unilaterally lead, bankroll, and ignore the free-rider effect of all the other nations.[27] This is important due to the fact that the United States needs the coalition mainly for moral support. The military, economic, intelligence, and political aid that is promised by these nations is helpful, and in some cases crucial, but it will also show that these issues are a concern to the other nations as well, not just the United States.

OEF Objective

So what is the objective of the OEF coalition? Or more importantly, what is the emphasis with respect to IO? Primary, most analysts agree that there are two key goals, the first being to defeat the terrorist organization al Qaeda and second to rehabilitate Afghanistan. Of course these requirements immediately require secondary or implied goals of winning the war on the ground in Afghanistan, and defeating the Taliban, who have been protecting Osama bin Laden. However according to the White House and State Department, the main focus of OEF should be, as quoted from their web page, ". . . that the Bush administration believes that 'there is no silver bullet, no single event or action that is going to suddenly make the threat of terrorism disappear . . .' " The main characteristics of this new war on terrorism will thus include the following:

- A broad-based and sustained effort—diplomatic, financial, intelligence, and military, both overt and covert—will continue until terrorism is rooted out.

- A situation similar to the Cold War, when continuous pressure from many nations caused communism to collapse from within.

- The United States will press the fight as long as it takes.[28]

All of the military operations to date are what many analysts consider as only the first phase of military operations. Scholars predicted that the next phase after OED would be follow-up missions in other nations of the region. This was based on statements by prominent Bush administration personnel such as Paul Wolfowitz, the deputy secretary of defense, who was quoted as saying that Saddam Hussein is ultimately a target as well, and, as predicted, the United States quickly lost most of its coalition members. It is highly unlikely that the White House will be able to maintain its tremendous amount of military, political, and financial support for a war on terrorism that has no end game, no moment of victory for all to savor.

Yet there is also a highly active intelligence, judicial, and financial aspect to the search for the terrorists as well. Unlike Operation Desert Storm (ODS), OEF is fully integrated as an interagency operation, with organizations like the Department of Justice, Health and Human Services, and Treasury Department all playing a leading role at the cabinet level, and below that, the FBI and the CIA are instrumental as well. For example, a key component to the successful conclusion of this mission will be if certain leaders of the different al Qaeda terrorist cells and organizations were captured or detained for arrest. The U.S. military is prohibited by law (*Posse Comitatus,* 1866) from conducting domestic legal actions, so it is imperative to obtain good horizontal communication between these different interagency groups, such as the FBI, to arrest these terrorists. That is where the role of intelligence plays such a crucial role.

As mentioned earlier, much of the efforts to stop the al Qaeda terrorist organization by the Bush administration are legal and financial in nature. Within ten weeks of the terrorist attacks on the World Trade Center, more than 350 people were arrested or detained around the world by a number of different foreign intelligence and police organizations. This is in addition to over 1,000 people that are currently under investigation within the United States by the FBI.[29] This immense partnership of law-enforcement agencies and intelligence activities is without precedent by not only the American organizations but by their foreign counterparts as well. Only under the umbrella of an anti-terrorist coalition, and the profound change in international politics that has occurred in the last few years, has such cooperation been possible.

Therefore, from a legal and judicial standpoint the coalition that originated in OEF is very different from ODS. In the former operation, these intelligence organizations were not playing such an open and public role in the overall conduct of the mission. However, that is not the case today. The Bush administration has aggressively undertaken a systematic isolation of the Taliban regime and al Qaeda organization by rounding up as many of the supporters as possible. It has been reported that the White House and the CIA have shared sensitive information with foreign nations to aid in the capture of suspects. Some of these arrests have foiled follow-up terrorist attacks, and others have been useful for leading police departments around the globe to more leads in their search for clues to locate Osama bin Laden.[30] In fact, some newspaper and magazine reports on these arrests have highlighted the increased sharing of information between the different intelligence services. Egypt, Pakistan, Jordan, Spain, Bahrain, Saudi Arabia, and Russia, in particular, have been very active and helpful to the CIA in their manhunt efforts. As noted in a magazine article, King Hussein of Jordan warned over ten years ago of bin Laden's burgeoning network, and his son, King Abdullah, has increased ties with the CIA since September 11th. Analysts speculated that Jordanian operatives were trying to infiltrate al Qaeda in an attempt to gain sorely needed intelligence for the United States.[31] Undoubtedly, one of the best lessons learned and most useful byproducts of OEF will be this increased cooperation between the different intelligence agencies. This includes not only the thirteen domestic activities that make up the U.S. intelligence services but also across the board with other key nations as the level of sharing between these organizations has risen dramatically. According to some, in the fall of 2001, the FBI had more than 25 percent of its agents (over 7,000 personnel) involved in thirty-five task forces concentrating on anti-terrorism operations.[32] Combine this with enhanced detention efforts by the Immigration and Naturalization Service and you see a tremendous amount of synergy developing from this coalition.

In addition, the financial network that supported and funded these activities is being restricted as well. With the Treasury Department in the lead, other cabinet-level agencies are helping as well, such as the Customs Service. By Christmas of 2001, the United States had blocked more than $27.7 million dollars in assets belonging to the Taliban and al Qaeda ter-rorist network, while OEF allies had blocked another $30.8 million. A total of 137 of the 195 nations in the coalition joined the United States in putting restrictions on these accounts, which number over 1,100 alone in the United States. This equates to sixty-two different organizations or individuals that are associated with the two money laundering networks of Al Barakaat and Al Taqwa that have all been in essence shut down by the U.S. Treasury Department.[33] Combine this with an action plan by the governments of 20 (G-20) to deny terrorists or their associates access to the financial systems and you have an entirely new cooperative environ-ment that did not exist a mere three months before.[34]

Likewise, on the diplomatic front the State Department led by Secre-tary Colin Powell was very aggressive in joining new members to the OEF coalition. They worked overtime, conducting a number of bilateral nego-tiations all in the interest of weaving the different nations ever closer together into an effective alliance. It is natural that this agency would have the lead on foreign policy issues, but the fact that they have enacted a number of key political issues is rather ingenious on the part of these gov-ernment officials. Some of these issues are:

- Enacting a U.N. Security Council binding resolution requiring all member nations to pursue terrorists.
- Invoking Article V by all nineteen nations of NATO declaring that an attack on one is an attack on all.
- Invoking the Rio Treaty by the Organization of American States (OAS) obligating all signatories to consider an attack against any member as an attack against all.

Not only is this the first time that these treaties have been invoked, but most academics would never have believed that if Article V were actually used that it would be for the United States instead of a European nation.

In addition, the State Department also promulgated its official counterterrorism policy, which in essence allows no concessions to terrorists and requires that they be brought to trial for their crimes.

Unfortunately, a sad fact of comparisons between the two coalitions may be the fact that both could, in the end, be extremely long lasting. Until the onset of Operation Iraqi Freedom (OIF), ODS was the longest-lasting military conflict for the United States in the twentieth century. American forces have been engaged against Iraq for longer than the active hostilities portion of the Vietnam War, and, with no end in sight, one has to ask if OEF will suffer the same fate. Right now the fight against terrorism is open ended, with troops still in Afghanistan, long after the Taliban were defeated. This correlation between the two coalitions is perhaps inevitable in this new era of the post Cold War conflict.

A Comparison of IO in OEF and Operation Iraqi Freedom (OIF)

On September 11, 2001 nineteen mostly Saudi Arabian followers of the Saudi dissident Osama bin Laden conducted a series of attacks against the United States that radically altered the American worldview. We may never know their complete target list. However, the three known targets were carefully chosen symbols of U.S. national power. The World Trade Center (WTC) complex symbolized the nation's global economic prowess. The Pentagon is recognized worldwide as the symbol of U.S. military might. The U.S. Capitol Building is the symbolic center of the U.S. political system; only this building was spared from attack, apparently by the heroic efforts of individual citizens who took charge of events on a hijacked airliner after learning of the attacks on the WTC and the Pentagon. Twenty-six days later, on October 7, 2003, the United States, leading an international coalition that grew to over fifty nations, invaded Afghanistan in the initial phase of what President George W. Bush called a "global war on terrorism" (GWOT).[35] The invasion was a direct response to the September 11th attacks. Operation Enduring Freedom (OEF) aimed to eliminate bin Laden's al Qaeda terrorist group at its primary operating base, Afghanistan. At the time, Afghanistan was a country ruled by the radical Islamist Taliban regime and the host to numerous training camps for terrorists from around the world. Its torturous terrain, paucity of economic infrastructure, and reputation for having defeated the invading Soviet Army in a bloody, protracted guerrilla war promised to make Afghanistan a difficult battlefield.

In March 2003, twelve years after winning Operation Desert Storm, a U.S.-led coalition began the second major phase of the GWOT, invading

Iraq. Iraqi dictator Saddam Hussein had long stood accused of aiding and abetting terrorists and had defied the United Nations since the end of Desert Storm by refusing to comply with the surrender terms. Though the United States could make no direct connection between the Iraqi regime and the attacks of September 11, 2001, Iraq's long history of harboring terrorists and its pursuit of weapons of mass destruction were deemed sufficient to warrant a regime change. The U.S. Department of Defense (DoD) christened the second Gulf war "Operation Iraqi Freedom" (OIF).

This section compares the United States' employment of IO in OEF and OIF and discusses the IO lessons learned from these operations. Though the focus of the discussion is on IO in the DoD, the chapter also addresses the strategic communications USG. Strategic communications are the media themes and messages originating from the executive branch of the USG, particularly the White House and Department of State. All information used in preparing this chapter comes from open sources—no classified information was used.

Looking back, in Operation Desert Storm the art and science of command and control warfare was applied to near perfection.[36] Command and control warfare, or C2W, targets an adversary's command and control (C2) systems. Military commanders use C2 systems to coordinate and synchronize virtually every aspect of their planning and operations. Applying C2W, the Allied Coalition in Desert Storm systematically bombarded Iraq's frontline troops with psychological operations (PSYOPS); crippled Iraq's integrated air defenses; blinded its target acquisition; shut down its propaganda machine; and totally disrupted military communications from the strategic to the tactical levels. These actions were supported by the most massive military deception operation since the Normandy invasion, which featured General Norman Schwarzkopf's famous "left hook," an operational maneuver that moved massed armor deep into Iraq to destroy Saddam Hussein's vaunted Republican Guard. Iraq's military, psychologically exhausted and rendered electronically deafened and blinded by C2W, was subsequently defeated by relentless aerial attacks combined with a lightning-swift ground campaign. The image of defeated Iraqi forces surrendering by the thousands remains vivid today. C2W was the genesis of the joint information operations (IO) doctrine employed by the U.S. military today.

Despite the U.S. military's enthusiasm for IO and its having had over five years to exercise its joint IO doctrine, at the time of this writing the DoD and military services have only recently almost agreed on a common definition of IO. The proposed definition (still not quite final) is, "the integrated employment of the core capabilities of electronic warfare, computer network operations, psychological operations, military deception, and operations security, in concert with specified supporting and related capabilities to influence, disrupt, corrupt, or usurp adversary human and automated decision-making while protecting our own."[37]

The discussion in this chapter divides IO into three categories, based upon an emerging U.S. Air Force IO doctrinal framework that is a convenient and logical way to "package" IO: influence operations, electronic combat operations, and network combat operations. Influence operations is the integrated planning and employment of military capabilities to achieve desired effects across the cognitive battlespace in support of operational objectives. Psychological operations (PSYOPS), military deception (MD), operations security (OpSec), counterintelligence (CI), and public affairs (PA) are elements of influence operations.[38] Electronic combat (EC) operations is the integrated planning and employment of military capabilities to achieve desired effects across the electromagnetic battlespace in support of operational objectives. Electronic attack (EA), electronic protection (EP), and non-lethal suppression of enemy air defenses (SEAD) are operational elements of EC operations. Network combat (NC) operations is the integrated planning and employment of military capabilities to achieve desired effects across the digital battlespace in support of operational objectives. Network attack and network defense are operational elements of NC operations.

The circumstances surrounding the beginnings of OEF and OIF were considerably different. This affected the way IO was employed during each. The manner of employment of IO in each operation also reflects the complexity, or lack thereof, of the communications infrastructures, both technical and human, in each area of operation. IO in OEF was dominated by influence operations, whereas IO in OIF was a more balanced mix of influence operations, EC operations, and NC operations. With the targets of September 11th still smoldering, the United States enjoyed considerable global support for invading Afghanistan. By the beginning of the second gulf war, a group of former allies, led by France and Germany, had blocked support in the United Nations Security Council for a U.S. invasion of Iraq.

OEF was begun with virtually "no-notice," meaning the Pentagon had only a minimal amount of time in the wake of September 11th to prepare the information battlespace prior to the initiation of hostilities. Consequently, IO suffered from a lack of sufficient planning and preparatory time. Afghanistan lacked a robust national communications infrastructure, making it impossible to inform the Afghan citizens of Coalition intentions using an information campaign conducted through the international media. The Coalition did not have sufficient time to employ a lengthy PSYOPS campaign before beginning hostilities, as would normally have been desired. The high rate of illiteracy within the Afghan population increased the difficulty of producing effective, printed PSYOPS products.

Part of the problem is that the Afghan population exists in a loose confederation of clans led by feudal warlords. In the beginning of the conflict, most did not even understand why the Americans had invaded. "Information warfare experts look for what they call 'the voilà moment.' In Afghanistan, the biggest lesson we learned in our tactical information operations—the radio and TV broadcasts—was the importance in explaining, 'Why are we here?' a senior American military officer said. 'The majority of Afghanis did not know that September 11th occurred. They didn't even know of our great tragedy."[39] The "voilà moment" in OEF wouldn't come until the Afghan population understood why the Americans and their Coalition partners had come to fight the Taliban. IO was the primary means of providing an explanation.

The situation at the beginning of OIF was considerably different. For all practical purposes, the United States and the United Kingdom had been at war with Iraq since 1991, flying daily combat missions to enforce the no-fly zones over Iraq in Operation Southern Watch and Operation Northern Watch. Iraq's well-educated, fairly sophisticated population boasted high literacy, a taste for Western media and entertainment, and enjoyed a robust national communications infrastructure. This included widespread access to global radio and television broadcasts via satellite. Women were well integrated into the machinery of the Iraqi government and society. Iraq was a far more fertile ground for waging IO than was Afghanistan. The situation in each country dictated the very different manners in which Coalition forces employed IO in the two main foreign playing fields of the GWOT.

Influence Operations During Operation Enduring Freedom

For countries that could never hope to match the United States militarily, IO has leveled the playing field somewhat. Even when a country lacks good communications infrastructure, like Afghanistan, the international media is a readily available medium for transmitting propaganda, whether wittingly or unwittingly. It is clear that Afghanistan's Taliban leadership manipulated the media to spread propaganda aimed at turning world opinion against Coalition operations. The tactics used had been seen before in the Balkans: intentionally damaging mosques to make it look as if they had been damaged by Coalition bombing; taking the press on well-staged hospital tours to see supposed victims of Coalition atrocities; and placing military vehicles in locations near mosques or other public buildings in the hope that collateral effects of an attack on the vehicle might damage the mosque or building, leaving incriminating "proof" of war crimes.[40] USG officials held several press briefings during OEF and OIF, specifically for the purpose of exposing enemy denial and deception (D&D) methodology and explaining in detail how they could prove that the purported evidence was faked. Most of the evidence came from comparing overhead imagery of the target areas before and after the attacks. This approach was fairly effective in neutralizing some of the enemy propaganda that might have otherwise required much greater effort to overcome.

The influence of the Qatar-based Al Jazeera broadcasting network, which has been called the "Arab CNN," cannot be overstated. Al Jazeera beams its message across the Middle East and is reported to reach 35 to 40 million viewers.[41] It stands accused by some of being a mouthpiece for Osama bin Laden, periodically airing recorded messages from bin Laden that may be used to pass cryptic messages to his followers. Regardless of its motives, Al Jazeera is a market competitor for U.S. PSYOPS and public affairs broadcasts in the Middle East.

The first such bin Laden tape broadcast by Al Jazeera demonstrated just how much confusion remains in the USG regarding IO, especially when it comes to harmonizing media operations. "When the first bin Laden tape was broadcast on Al Jazeera television, followed by governmental requests to commercial stations not to rebroadcast for fear it may contain coded messages to terrorist 'sleepers', Voice of America ignored the request."[42] For some reason, the USG-operated Voice of America failed to comply with the request, even as some commercial media outlets suppressed it. The fact that the USG asked commercial media to suppress the tapes was used as ammunition by the enemy to accuse the "free-speech loving" Americans of hypocrisy.

Al Jazeera was hard at work during OEF, broadcasting messages from Osama bin Laden even as U.S. Army troops searched for him amongst the caves in the mountainous terrain along the border between Afghanistan and Pakistan, where bin Laden was believed to be hiding. Al Jazeera also aired daily interviews with Afghanistan's ambassador to Pakistan. The ambassador, a member of the Taliban regime, used these interviews to spread misinformation about alleged civilian casualties resulting from American combat operations in Afghanistan. This one-sided coverage proved troublesome for the U.S. leadership, who had difficulty deflecting international criticism of their operations that grew steadily with the daily interviews of the ambassador.

The United States made limited use of television in an attempt to explain the September 11th attacks to the Afghan population. Few people in Afghanistan own televisions, so broadcasting a television message would have been of little value. Instead the Americans produced a three-minute video explaining the attacks and promoting support of the government of President Hamid Karzai. The video was then shown around the country through a series of visits with local officials and citizens by U.S. PSYOPS troops.[43] Some individuals who saw the video obviously understood. Others, some of which had never even seen television before and who were unfamiliar with the images of a big city and skyscrapers, seemed not to understand. The overall effectiveness of this methodology cannot be accurately assessed, but under the circumstances it was probably the best that could be accomplished using televised media. One of the key requirements for effective PSYOPS is selecting the proper media and products to suit the target audience. Television was not the media of choice for broadcasting PSYOPS messages in Afghanistan.

As in most recent conflicts, PSYOPS leaflets were a key element of U.S. influence operations in OEF. On November 8, only a month into the operation, the Pentagon announced that 16 million PSYOPS leaflets had been dropped on Afghanistan.[44] Because of the Afghan population's high illiteracy rate, many leaflets were cartoon style, with little or no writing. The challenge of developing effective leaflets of this type is that different people will interpret them in different ways. This led to at least one situation with potentially deadly results. U.S. aircraft dropped humanitarian daily rations (HDR) in many areas of Afghanistan. The HDR, packaged in bright yellow plastic, are prepared in accordance with the strict dietary requirements of the Koran. Unfortunately, the HDR packets are similar in size and color to the sub-munitions in certain cluster bombs. The aircraft dropping HDR also dropped cartoon-style leaflets showing an Afghan-looking family eating the rations. When someone realized that the leaflets might lead some to pick up the yellow sub-munitions, the Pentagon had to scramble to correct the problem. They quickly produced a new leaflet warning the Afghan population not to pick up the sub-munitions that looked like HDR. Human rights groups had criticized the use of the cluster bombs because it was thought that children might be attracted to the bright yellow color of the sub-munitions. The Pentagon reported that the color of the humanitarian daily rations would be blue in the future.[45]

The Taliban regime published propaganda pamphlets aimed at discrediting Coalition forces. Offering monetary rewards for killing or capturing Westerners and threatening retaliation for cooperating with them, the pamphlets were slipped under peoples' doors during the night. This led to their being nick-named "night letters."[46] Similar pamphlets were also circulated amongst refugees in Afghanistan and Pakistan. The Taliban propaganda threatening violence against their own people stands in stark contrast to the U.S. PSYOPS and media methodology. There was no apparent attempt by the Taliban to target U.S. or coalition forces with propaganda, something that has proven fruitless in previous conflicts.

Electronic Combat Operations During Operation Enduring Freedom

The simple description of electronic combat (EC) operations in Afghanistan is that most of the limited Afghan communications architecture and air defenses were taken out by physical and/or electronic attack during the initial hours of the coalition air attacks. The electronic warfare and air defense environment that coalition aircrews faced in Afghanistan can be described simply as low-threat or non-hostile. Due to this low-threat air defense environment, the U.S. Navy's EA-6B jamming aircraft, traditionally used for suppression of enemy air defenses (SEAD), were employed to ". . . exploit new techniques to jam ground communications by working with the EC-130 and other electronic intelligence gathering aircraft."[47] The DoD displayed similar innovation during OIF, when the U.S. Air Force for the first time used the *Compass Call* electronic attack aircraft to broadcast PSYOPS messages with its powerful transmitters.[48] Such innovation exemplifies the synergy that can be gained through combining the effects of IO capabilities.

The following paragraphs, though grouped under the discussion of electronic combat operations, could equally have been placed under the discussion of influence operations. This is typical of so many information operations. For example, bombing an enemy radio station to silence its propaganda broadcasts is considered a form of IO, according to current U.S. joint IO doctrine, because it is intended to thwart the enemy's influence operations. (The new definition of IO being considered by the DoD might not classify this act as IO.) Taking over an enemy frequency for your own uses would be considered electronic warfare in some circles. So whether the following examples constitute influence operations or EC operations, the lessons to be learned are the same.

Before OEF began, "the Taliban-run Voice of Sharia radio, broadcast from Kabul, filled the airwaves with religious discourse and official decrees. Their opponents, listeners were told, were 'evil and corrupt forces.' Today, that Taliban signal has turned to static, its transmitter destroyed by two cruise missiles."[49] Having taken out the transmitters used by the Taliban to spread propaganda about the coalition forces, the U.S. military took over the main frequency used for Taliban radio broadcasts and used it for their own PSYOPS broadcasts to the Afghan population.

They also distributed some 7,000 radios to the Afghan people.[50] Because of the isolation of many Afghan villages, it was difficult to reach them with either radio or television broadcasts. This problem was addressed using typical American innovation. "So Special Forces troops made contact with local coffee-house managers, and offered them the same radio programs being broadcast from *Commando Solo* planes, but on compact discs to be played over a boom box for the patrons. The program gave birth to a new icon on the military's maps of Afghanistan: a tiny picture of a coffee mug to indicate the location of village businesses that agreed to play CD copies of the American radio programming."[51] The EC-130E *Commando Solo* is a cargo plane converted to conduct PSYOPS broadcasts and is discussed later in this chapter.

The war in Afghanistan saw the U.S. Army's *Prophet* unmanned aerial vehicle (UAV) tested operationally for the first time. The initial feedback from users was extremely favorable. The *Prophet* is a multi-mission platform that provides ground force commanders with the capability to conduct tactical signals intelligence (SigInt) operations, measurement and signatures intelligence (MasInt) operations, and electronic attack operations.[52] The *Prophet* employment is part of a trend of increasing use of UAVs in U.S. military operations. The multi-mission nature of UAVs make them prime candidates for answering some of the IO requirements of the future, including electronic attack, PSYOPS broadcast, ComSec monitoring, OpSec evaluation, computer network attack, public affairs and civil affairs broadcast, and a host of other applications. As U.S. military technology evolves, the distinction between intelligence, surveillance, and reconnaissance (ISR) systems and IO systems will continue to blur.

In the post-Taliban Afghanistan, influence operations continue to be the focus of DoD IO. In particular, civil affairs and media operations are crucial. At the one-year anniversary of the beginning of OEF, the coalition had improved Afghanistan's infrastructure to levels never imagined before the war. The statistics are impressive:

- Ten water projects were completed during the first six months of 2002. These included 83 wells, benefiting approximately 260,000 Afghans. An additional 16 new water projects had been approved.
- De-mining teams from Norway, Britain, Poland and Jordan helped clear mines from more than 1.8 million square meters of terrain.
- U.S. Army civil affairs troops completed 61 school repair projects with plans for 44 more.
- The U.S. provided 10 million textbooks and 4,000 teacher-training kits.
- Canada, Greece, Belgium and Iceland delivered 60 metric tons of goods.

- Jordan built a hospital in Mazar-e-Sharif that treated more than 105,000 patients.

- Belgium led a multinational humanitarian assistance mission that delivered 90 metric tons of UNIMIX to starving children in Afghanistan.

- The U.S. jointly funded the measles vaccinations of more than four million children.[53]

Ultimate success in Afghanistan will depend heavily upon the success of humanitarian relief operations. It is an indisputable fact of military operations that returning a sense of normalcy to the citizens of a vanquished nation is one of the surest ways to stabilize the country. It is also critical that media operations communicate the details of humanitarian relief operations to the population. It is for this reason that U.S. Army civil affairs units have an embedded public affairs capability.

Network Combat Operations During Operation Enduring Freedom

Although the details of any DoD network combat (NC) operations employed during OEF remain classified, it is safe to assume that they had little, if any, impact on the operation. Afghanistan was a technologically unsophisticated adversary that posed no significant threat of computer network attack and likewise possessed little network infrastructure that would warrant a sophisticated CNA. Iraq would prove a much more fertile ground for NC operations.

Influence Operations During Operation Iraqi Freedom

During the build-up to OIF and since, OpSec has been a constant concern for the USG and the DoD. U.S. Secretary of Defense Rumsfeld issued new website OpSec guidance to the DoD on January 14, 2003, noting that an al Qaeda training manual recovered in Afghanistan states, "using public sources openly and without resorting to illegal means, it is possible to gather at least 80 percent of information about the enemy."[54] Before the September 11th attacks, it was possible to access a vast amount of operationally sensitive information from government, military, and commercial websites. Detailed maps of nuclear power generation plants, DoD military installations, and key public buildings such as the U.S. Capitol are just a few examples of information that was freely available to any terrorist with Internet access. It is now clear that all levels of government activities, the DoD, and key private organizations comprising the United States' critical economic infrastructures must have an OpSec review process to review information before it is placed on a publicly accessible website.

The Bush administration received considerable criticism for suppressing the news media during OEF. As *New York Times* journalist Elizabeth Becker reported on the war in Afghanistan, the "military has imposed a tight lid . . . trying to walk the fine line of saying enough to reassure the public that the war is on target but keeping the news media at bay."[55] The so-called act of "embedding" journalists with army and marine combat units during OIF was an answer, at least in part, to such criticism.

Embedded journalists enabled the world to see the "ground truth" in Iraq, which somewhat aided in neutralizing Iraq's propaganda efforts. However, the embedded journalist sometimes exacerbated the OpSec problem. Reporter Geraldo Rivera was expelled from the 101st Airborne Division's zone of operation because of an OpSec violation. "Rivera violated the cardinal rule of war reporting by giving away crucial details of military plans during a Fox News broadcast from Iraq. During the broadcast, Rivera asked his photographer to aim the camera at the sand in front of him. He bent down and drew a map of Iraq in the sand showing the comparative location of Baghdad to his unit. Rivera even proceeded to draw diagrams of where the unit was heading next. If enemy forces were watching the news broadcast, they could have launched a preemptive strike against the whole Airborne Division."[56] Rivera later admitted that he had made a mistake and apologized for the blunder. In the past, the media has posed little immediate threat to combat forces. However, since the advent of global television and radio broadcast from the frontlines, OpSec has become a major issue. The 1st Marine Division Commander, Major General James M. Mattis characterized the embedded journalists concept as "a limited success."[57] His opinion of the journalists is not as positive as that expressed by Secretary of Defense Rumsfeld or CentCom Commander General Tommy Franks. The jury is still out on whether to use embedded journalists during future U.S. military operations.

OpSec also became an issue on the home front. An egregious violation potentially put the family members of a B-1 bomber crew at risk during OIF. On April 7, a B-1 bomber crew conducting a mission over Iraq was diverted to a new target. The target was later revealed to be a building in which Saddam Hussein and some of his senior leadership were believed to be located. In a subsequent interview broadcast from the Pentagon, a number of crew members' full names, their commanding officer's name, their unit, and their home base location were identified.[58] Given the threat of terrorism in the U.S. today, this incident could have put the family members of these individuals at risk. With today's powerful Internet search engines, given a name and location, it is often possible to obtain an individual's phone number, email address, street address, and a map to their place of residence in a matter of minutes. Identifying the unit, base, and full names of the B-1 crew during a media event was an OpSec violation of the worst kind. Attacks on family members of deployed service men and women could have dire operational consequences for the U.S. military. Deputy Secretary of Defense Paul Wolfowitz issued a new OpSec directive to the DoD on June 6, 2003, directing the military services to reassess their OpSec programs by October 1, 2003, and providing new objectives for the assessment.

The U.S. military officially acknowledged the human factors element of influence operations when it revealed that senior Iraqi military leaders were bribed into submission before the onset of ground combat in Iraq. General Tommy Franks, the CentCom commander, indicated that senior Iraqi officers accepted bribes for a promise not to engage Coalition forces. Consequently, Coalition forces met light resistance in many locations that might have otherwise been heavily defended. "I had letters from Iraqi generals saying: 'I now work for you'," General Franks said.[59] This same type of influence was employed years earlier to encourage the peaceful departure into exile of General Raul Cedras during Operation Uphold Democracy in Haiti.[60]

During OIF, President George W. Bush employed national strategic communications to directly address Iraqi generals in a nationally televised speech during which he made it clear that, ". . . anyone ordering the use of weapons of mass destruction will be treated as a war criminal and likely will be executed."[61] As is often the case with influence operations, it is difficult to assess the effects of the President's warning. However, the large number of prisoners from OEF that were being held in a U.S. military prison in Guantanamo Bay, Cuba were ample evidence of the veracity of his words and may have contributed to the psychological impact and effectiveness of the warning. The DoD has yet to develop adequate measures of effectiveness (MOE) for evaluating influence operations and it may prove an impossible task, given that human behavior is so unpredictable.

As usual, PSYOPS broadcasts from the Air Force's EC-130E *Commando Solo* aircraft were a key element of the OIF information campaign. One program broadcast by the EC-130E mimicked a popular Iraqi radio program "Voice of Youth." "The American programs opened with greetings in Arabic, followed by Euro-pop and 1980's American rock music— intended to appeal to younger Iraqi troops, perceived by officials as the ones most likely to lay down their arms. The broadcasts included traditional Iraqi folk music, so as not to alienate other listeners, and a news program in Arabic prepared by army psychological operations experts at Fort Bragg,

North Carolina. Then comes the official message: "Any war is not against the Iraqi people, but is to disarm Mr. Hussein and end his government."[62]

The inability of *Commando Solo* to operate in areas of mid- to high-intensity air defense threat has long been a problem, as has the limited range of its television and radio broadcasts. A Defense Science Board study published in 2000 indicated that the small, aging fleet of EC-130Es is no longer capable of meeting the PSYOPS needs of the CCs.[63] To help remedy this, during OIF Coalition forces employed PSYOPS broadcasts from ships operating in the Persian Gulf.[64] The U.S. Navy in recent years deployed a containerized PSYOPS broadcast package that can be placed on and operated from the deck of a ship. The U.S. Navy has also improved its ability to produce printed PSYOPS products while underway, rather than having to have them delivered to their ships. As the military services continue to develop new concepts and requirements for operational platforms and systems, PSYOPS broadcasts, leaflet delivery, and new means of delivering PSYOPS messages should be considered as a potential capability for each.

During OIF, the Al Jazeera network was once again at work. It received considerable criticism from American officials for broadcasting video of Americans captured by the Iraqis, including one that appeared to show that some of the prisoners had been summarily executed. Many Americans citizens cried foul and accused Al Jazeera of outright support of Saddam Hussein. The USG's response to Al Jazeera thus far has been rather clumsy. The government has yet to display a coordinated strategic communications strategy to deal with Al Jazeera and similar Middle Eastern media outlets. Until this happens, the United States will continue to lose ground in global media confrontations.

OIF did have its lighter moments. Mohamed Saaed Al-Sahhaf, the Iraqi Information Minister who is known today by his nickname "Baghdad Bob," became a celebrity through parodies of him on the "David Letterman Show," the "Tonight Show," and a website dedicated to his infamous railings (www.welovetheiraqiinformationminister.com). Al-Sahhaf claimed in his now famous taped interviews that the Iraqi armed forces were slaughtering Coalition forces, even as Coalition tanks drove through the streets of Baghdad and Coalition forces occupied Saddam Hussein's palaces. The lesson for anyone who watched "Baghdad Bob" is that truth is the strongest weapon in PSYOPS. Unlike the propaganda employed by most of its adversaries, U.S. PSYOPS doctrine focuses on transmitting selected, truthful messages to specific human targets in order to influence their actions. "Baghdad Bob" is an excellent example of how not to conduct PSYOPS. "I triple guarantee it!"[65]

As in Operation Desert Storm, Iraqi denial and deception operations during OIF were effective. During Desert Storm the Coalition devoted considerable time and resources to locating and destroying Iraq's elusive mobile SCUD missile launchers, with no success. Some claim that Iraq's interaction with the United Nations during the twelve ensuing years after the first Gulf War was one massive deception operation. The certainty with which President Bush and Great Britain's Prime Minister Tony Blair accused Iraq of possessing weapons of mass destruction (WMD) and the inability of the Coalition forces now occupying Iraq to locate a single WMD suggest that Iraq may have pulled off one of the greatest deception operations in our time.

It is now clear that the Iraqis employed a number of innovative means to deny and deceive Coalition intelligence surveillance and reconnaissance during OIF. These included placing weapons facilities inside civilian neighborhoods, faking collateral damage for the sake of exploiting the media, and hiding weapons and munitions inside schools, mosques, and walls of buildings.[66] Scores of official documents have been recovered from private homes. Perhaps the most innovative D&D method was burying equipment and documents. The Army's 101st Airborne Division discovered eleven mobile laboratories, suspected of being biological agent production labs, buried near the town of Karbala.[67] "Officials said they have found tons of military equipment, including airplanes, buried beneath the sand, and they believe illegal weapons and the laboratories to make them will have been hidden in such a manner."[68] Finding buried materials in a country the size of California is like searching for a needle in a haystack. Much of the equipment and documents discovered so far were located as a result of tips provided by Iraqi citizens. Despite the U.S. military's sophisticated ISR capabilities, it is painfully clear that there is still no substitute for good human intelligence on the ground in an area of operations. However, sophisticated ISR proved useful for exposing Iraqi D&D techniques to the media, as was clearly demonstrated in a State Department media briefing on October 11, 2002. The briefing was dedicated solely to explain to the media the D&D methods used by Iraq during OIF.[69] D&D methods will only improve with time, meaning future ISR sensors will have to defeat increasingly sophisticated D&D methodology.

Electronic Combat Operations During Operation Iraqi Freedom

Suppression of enemy air defenses (SEAD) remains critical to the U.S. Air Force's ability to accomplish its mission. One Coalition fixed-wing aircraft was lost to Iraqi air defenses during OIF. Another was downed by friendly fire. A software flaw in the Army's patriot missile fire control system may have caused the friendly fire incident.[70] One of the key weapons in the SEAD effort is the AGM-88 HARM missile. The HARM is an anti-radiation missile employed against radar emitters. It is usually associated with the Navy's EA-6B Prowler aircraft. Serb Air Defense units in Kosovo learned that they could avoid the missiles by turning off the radar emitters after handing a target off to a missile launcher.[71] A recent software upgrade to the AGM-88 saw its first combat employment in OIF. The Block V and Block IIIa upgrades maintain their targets in memory after the targeted radar emitter is switched off. The upgrade also has the option to "home-in" on emitters, attempting to jam the global positioning system (GPS) receivers.[72] The AGM-88 upgrade proved effective in OIF. This may account for why so many Iraqi radar sites never turned on their radars at all, although it cannot be proven that this was the cause. The improvements to the AGM-88 target memory are a good example of how the DoD and industry need to respond quickly when sophisticated adversaries develop D&D measures against U.S. technologies.

Coalition electronic deception was effective during the OIF air war. The U.S. Navy employed the tactical air-launched decoy (TALD) system against Iraqi air defenses. The TALD is a small, air-launched glider that flies a pre-programmed course for about 100 kilometers. The Navy found that during night attacks in particular, Iraqi air defense guns and missiles often fired at the TALD decoys, reducing the risk to combat aircraft. As a result, the Navy has submitted orders for an improved TALD (ITALD) that has a small jet engine.[73] As adversaries develop greater air defense technologies, it will be increasingly necessary for the DoD and industry to develop electronic D&D capabilities to protect U.S. aircraft.

By the time the second gulf war began, Iraqi forces had reconstituted their air defenses and could boast a robust, if somewhat dated, integrated air defense system (IADS). Before the war, U.S. military planners expressed concern over Iraq's acquisition of GPS jammers from Russia. This was due to the U.S. military's growing reliance on GPS navigation for precision weapons, particularly the GBU 31/32 Joint Direct Attack Munitions (JDAM). Their concerns appear to have been unfounded. In a press conference on March 25, 2003, Major General Victor Renuart, the USCentCom operations officer (J-3) confirmed that the Iraqis had employed six GPS jammers during the previous two nights of operations and that all had been destroyed. He added that one of the Iraqi GPS jammers had even been destroyed by a GPS-guided precision munition.[74] It appears that the DoD's heavy reliance on GPS may not have created as great a vulnerability as originally thought. However, greater numbers of GPS jammers employed by a more sophisticated opponent might have proven more effective. OIF simply did not provide enough data from which to draw firm conclusions on this issue.

Network Combat Operations During Operation Iraqi Freedom

Although the details of sophisticated network attack operations are classified, the DoD did confirm that the United States employed a barrage of e-mails, faxes, and cell phone calls to numerous Iraqi leaders in an attempt to persuade them not to support Saddam Hussein.[75] As might have been expected, these technical NC operations were conducted in support of PSYOPS. In the Information Age, e-mail addresses, fax numbers, and cell phone numbers can be a valuable tool to an adversary, as can the contents of a Microsoft Outlook or other electronic address book. Collecting this type of information will become increasingly important for the U.S. intelligence community and the intelligence services of U.S. adversaries.

This new type of highly personalized influence operations, supported by computer network technology, emphasizes the importance of employing good OpSec to protect sensitive but unclassified information. It also emphasizes the importance of having up-to-date anti-virus software and training computer users to delete e-mails from individuals they do not know. This is because many Trojan horse programs, which are often transmitted as attachments to e-mails, exploit the contents of address books to mail themselves to everyone on the address book once the e-mail attachment is executed. Malicious code has become increasingly sophisticated to the point that it is now possible to execute malicious code by simply opening an e-mail message, without having to open an attachment.

It is difficult to derive lessons on computer network defense (CND) from OIF. There was a slight increase in cyber-attacks against DoD systems during the initial days of OIF, but nothing significant was reported. It is not known whether these "attacks" were simple scanning or attempted network penetrations. The bottom line is that there was "nothing systemic that could be tied back to the enemy."[76] The military should not take comfort in the fact that so far during OIF there have been no known penetrations of DoD systems attributable to the enemy—the Iraqis never possessed a sophisticated network attack capability. Richard Clarke, former chairman of the President's Critical Infrastructure Board, recently issued a stern warning stating, "IT [information technology] has always been a major interest of al Qaeda. We know that from the laptops we have . . . that we've recovered that have hacking tools on them . . . it is a huge mistake to think that al-Qaeda isn't technologically sophisticated, a fatal one. They are well-trained, they are smart. They proved it on 9/11 with one style of attack, and they can prove it again."[77]

OEF and OIF IO Summary

"To maintain information dominance, we must commit to improving our ability to influence target audiences and manipulate our adversary's information environment. Continued development of these capabilities is essential."

General Tommy Franks[78]

One of the easiest conclusions to draw from the employment of IO in the GWOT is that the ability to effectively command and control (C2) seamless IO, particularly at the strategic and operational levels, remains elusive. This is not for lack of effort, but simply reflects how difficult it is to plan and synchronize the myriad tasks that must be successfully performed in order to accomplish the objectives of an IO campaign. J. Michael Waller provides a wonderful summary of the challenge: "To date, most U.S. public-diplomacy and information operations in support of the war effort have been piecemeal, tactical and mostly reactive instead of strategic, comprehensive and anticipatory. A long-term strategy has yet to be developed, according to administration officials. That, critics say, leaves the enemy to define the terms of debate and severely complicates U.S. diplomacy and military planning."[79] Developing a comprehensive national IO strategy for the GWOT should be a top priority for the United States.

Coordinating IO is particularly difficult when it requires the cooperation and synchronization of activities between organizations in the interagency, which is so often the case. One example surfaces far too frequently, the seeming inability of the White House, Department of State, and DoD to harmonize the themes and messages in their media releases. Each has a robust media capability, yet each seems unable to effectively coordinate with the other. This frequently results in mixed messages that confuse friends and adversaries alike as to the plans and intentions of the United States. This is a major obstacle to convincing others to act in a manner that supports U.S. national security objectives. It may be the greatest IO problem the United States needs to solve. The media can be a strategic enabler in a number of ways: to communicate the objective and end-state to a global audience, to execute effective psychological operations (PSYOPS), to play a major role in deception of the enemy, and to supplement intelligence collection efforts.[80]

The failure of the Pentagon to successfully establish the Office of Strategic Influence (OSI) is another example of coordination difficulties. This particular situation was further complicated because it was rumored that one of the office's missions was to engage in so-called "black propaganda" operations. This is the term used to describe the surreptitious placing of false information in the foreign media in order to influence the behavior of adversaries. Part of the controversy over using black propaganda is that misinformation planted abroad can reach and influence American audiences via today's global communications. This goes against the grain of American culture and raises a number of legal issues as well. As J. Michael Waller observes, "the Department of Defense set up a new OSI to run information operations abroad in support of U.S. strategic-defense goals. However, OSI hardly had gotten off the ground when the bureaucracy and turf-jealous senior officials leaked misleading, inflammatory and utterly dishonest stories that falsely portrayed the OSI as intending to plant 'disinformation' in the (American) press."[81] The OSI fiasco was a blow to the United States' ability to conduct strategic PSYOPS and a blow to the credibility of U.S. strategic communications throughout the world.

The Chairman of the Joint Chiefs of Staff, General Richard B. Myers, summed up IO in Afghanistan as follows: "It took too much time to put together the team. We missed the opportunity to send the right message, sometimes we sent mixed signals, and we missed opportunities as well."[82] Obviously, coordination efforts, particularly with other government organizations in the so-called interagency, continue to hamper DoD IO. The C2 of IO during operations in Iraq appeared to be better coordinated. Apparently the IO lessons from OEF were applied with some success in OIF. This is particularly apparent when one compares the restrictive media policy applied by the DoD in Afghanistan with the embedded media employed in Iraq. The DoD has recently taken steps to improve the C2 of IO. A 2003 change to the UPC, the document that describes the roles and missions of the U.S. military's major CCs, created a Joint Force Headquarters for IO (JFHQ-IO) under the command of the U.S. StratCom. This change effectively consolidates the responsibility for DoD IO under StratCom. The effectiveness of the change has yet to be demonstrated, but in theory it should improve coordination of IO, at least within the DoD.

Another clear lesson is that the USG has yet to discern how to effectively deal with the profound psychological impact of terrorism, such as the September 11th attacks. The attacks, ". . . had deep human, economic, and psychological impacts. The terrorists were not deterred by our overwhelming military superiority, in fact, for that day at least, they made it irrelevant. Traditional concepts of security, threat, deterrence, warning and military superiority don't completely apply against this new strategic adversary."[83] According to the Chairman of the Joint Chiefs of Staff, "This is a new kind of war. The military may not be decisive."[84] These observations emphasize the criticality of improving coordination in the interagency environment.

Americans living in the post-September 11th era have a changed worldview—a change that occurred literally overnight. The false sense of security provided by the nation's relative geographic isolation for over two hundred years is forever departed. Americans now live in a changed society. In the wake of September 11th, the United States enjoyed tremendous international support for its military operations in Afghanistan. However, the Bush administration's new policy of preemptive engagement to strike potential enemies before they strike the United States, as was used in the invasion of Iraq, has eroded that support. The strategic communications efforts of the United States have failed to help garner international support for stabilization operations in Iraq. The Patriot Act was enacted to help law enforcement agencies root out terrorists operating in the United States, but many citizens believe that the act infringes upon their constitutional rights. Increased security at airports, public buildings, and public events has changed the daily routines of Americans. An anthrax scare nearly crippled the U.S. postal system and slowed government operations when contaminated letters were sent to members of Congress. A pair of deranged snipers created massive public fear and chaos in the greater Washington, D.C. area over the course of a month in September and October 2002. A major electric power failure in the northeast U.S. raised immediate fears of a terrorist attack. The dread of terrorism is now firmly fixed in the American psyche, so the terrorists have scored a major victory. How should America respond?

Part of the response must be the continuation of diplomatic, military, legal, intelligence, and technical efforts to cut off terrorist funding and logistics throughout the world. It is also essential that the USG develops a comprehensive national IO strategy for the GWOT and continues its efforts to develop an effective strategic communications policy that clearly states U.S. actions and clearly conveys U.S. intentions to friends and enemies alike. Though the USG continues to focus on the wider GWOT, military stability operations continue in Afghanistan and Iraq. President Bush recently declared that Iraq has become the central stage for the GWOT, due to the large number of terrorists that have entered the country to engage Coalition forces. It is essential for the DoD to quickly and thoroughly analyze the IO lessons learned and to begin developing solutions to the problems identified. DoD IO for the remainder of OEF and OIF will focus mainly on influence operations. It would therefore be wise to focus on analyzing influence operations lessons learned to ensure that mistakes in this critical area of IO are not repeated. Media operations, PSYOPS, and civil military operations will play a key role in the success or failure of these missions.

Summary

The four areas of CNA, space, EW, and perception management are the largest growth areas for offensive IO in the last three years. However, that said, since September 11, 2001, counterterrorism and homeland security may eventually dwarf these as well. There are still many challenges to the successful application of IO for offensive operations. Because military planners organize their staffs along operational lines and continue to think of military operations as either offensive or defensive, information operations are causing many to challenge these traditional conceptions. The need for increased integration and cooperation among the diverse members of the interagency community, as well as the private sector, academia, and others, will eventually force those within the DoD to come to terms with the limitations imposed by traditional military planning methods and procedures. The nation which best develops a coherent national security strategy and thoroughly integrates both offensive and defensive information operations into all aspects of its diplomatic, informational, military, and economic policy will be best positioned to gain information superiority. In pluralistic democratic societies like the United States, the ability to develop such an approach may be illusive, given the many competing interests of all the players. Only time will tell how successful the United States will be in utilizing the offensive capabilities of IO in the future.

CHAPTER 5

Related and Supporting Activities

Organize, Train, and Equip

If a man does not know to what port he is steering, no wind is favorable.

Seneca

By law, specifically Title X of the United States Code, the DoD is required to organize, train, and equip a military force to protect the American people from adversaries both foreign and domestic. That interpretation has changed over the years, and in this chapter, the authors will show you how the different services have adapted to the revolution in warfare that is accompanying information operations. In addition, because the planning for warfare in the Information Age is so different from the normal planning cycle, in this section we will introduce you to these concepts as well.

IO Planning

Planning is the essence of military operations. Behind most military operations lie countless hours of planning and rehearsal. Military organizations at the tactical, operational, and strategic levels have organizational structures and detailed procedures for planning. These planning structures vary greatly in size and composition, but all have one thing in common: a desire to predict every possible outcome of an operation and to plan for every possibility. This chapter examines the utility of IO in support of peacetime engagement planning and in the deliberate and crisis action planning processes and discusses the DoD organization for IO planning.

Peacetime engagement is one of the principle missions of every regional Combatant Commander. The National Security Strategy (NSS) of the U.S. lists IO as one of the key capabilities required by the military.[1]

IO can support a Combatant Commander's peacetime engagement activities by helping to deter aggression and coercion, build coalitions, promote regional stability and promoting U.S. forces as role models for militaries in emerging democracies. The military accomplishes this mission through means such as forward stationing or deployment of forces, defense cooperation and security assistance, and training and exercises with allies and friends. Until fairly recently, the regional Combatant Commanders established their own direction for peacetime engagement in their respective theaters. This changed in 1997 when the CJCS issued planning guidance to the regional Combatant Commanders directing them to develop multi-year plans for theater peacetime engagement.[2] Originally these TEPs or Theater Security Cooperation Plan(s) (TSCP) were five years in length, but in 2000 they were lengthened to seven years.

The Chairman of the Joint Chiefs of Staff Manual (CJCSM) 3113.1, *Theater Engagement Planning,* published in February 1998, formalized the theater engagement planning guidance issued by the CJCS in 1997. This manual directs the regional Combatant Commanders to formulate their peacetime engagement strategies and to submit them for approval to the CJCS. Nowhere does CJCSM 3113.01 direct the Combatant Commanders to use an IO approach for peacetime engagement, but an IO strategy offers a logical means of accomplishing the Combatant Commanders peacetime engagement objectives. Because one of the principal attractions of IO is its potential to deter conflict, one would expect to see IO play a major role in the TSCP process. In fact, this is not so simple. Though a regional Combatant Commander may wield significant power in a theater or a country during a conflict, most of the authority lies in the hands of the DoS during peacetime. It is therefore incumbent upon the regional Combatant Commanders to coordinate with the State Department and a host of other USG agencies when planning peacetime engagement activities, especially ones involving the employment of IO. In particular, a regional Combatant Commander's staff may find themselves coordinating information operations with any of the following organizations:

- The Department of Homeland Security (counter-terrorism)
- The Drug Enforcement Agency (counter-narcotics)
- The FBI (counterintelligence)
- The DoE (nuclear weapons counter-proliferation)
- The DoC (foreign technology transfer)

Managing the details of an operations plan is becoming increasingly difficult, as technology has multiplied the number of information planners involved in the operations plan. The pathway for the U.S. military described in *The Concept for Future Joint Operations: Expanding Joint Vision 2010* describes how the United States must gain and maintain information superiority over its adversaries. Information superiority, as you may recall, consists of three components: information systems, information operations, and relevant information. Relevant information is all of the information of importance to the JFC in his exercise of joint command and control. It includes information about friendly forces, the enemy, and the operations area. Therefore it is incumbent upon the command to organize the IO cell so that it has the necessary balance of talent and expertise to sort through the profusion of information available to the average military staff today.

JP 3-13 provides some guidance to planners in the integration of IO in the planning process. The joint planning process is documented in the manuals of the joint operations planning and execution system (JOPES). To incorporate IO into military planning, the DoD has added a number of special units and organizations as mentioned in Chapter 1. At the highest level, the Joint Staff J3's Deputy Director for Information Operations (DDIO) provides guidance, direction, and support of IO planning at the unified commands. Likewise, the IO cell at a unified command performs similar functions for any sub-unified commands, joint task forces, and/or service component commands subordinate to it. In addition, as covered earlier in the first chapter, all services have formed component IO or IW centers to support all aspects of IW. Although each differs in organization and mission, all provide support to IO and IW planning. The following discussion examines the approach that each military service has taken towards IO and IW planning.

Military IW Service Centers

The 1st Information Operations Command (1 IOC) (previously named the Land Information Warfare Activity) at Fort Belvoir, Virginia supports deployed army units, both operationally and during exercises. 1 IOC teams support the army commander's goal of achieving information dominance with the other JTF components or organizations. 1 IOC's purpose is to provide army commands with technical expertise that is not resident on the command's general or special staff, and to exercise technical interfaces with other commands, service components, and national, DoD, and joint information centers. When deployed, 1 IOC field support teams (FST) become an integral part of the command's IO staff. To facilitate planning and execution of IO, 1 IOC provides operational support to land component and separate army commands, and reserve component commands as required. 1 IOC has had FSTs deployed in the Balkans since 1996 and has gleaned innumerable lessons learned.[3] The 1 IOC also provides the U.S. Army's computer emergency response team (CERT).

The Fleet Information Warfare Center (FIWC) at Naval Amphibious Base in Little Creek, Virginia supports IO and IW planning for the U.S. Navy and the U.S. Marine Corps.[4] FIWC provides naval and joint commanders with deliberate and crisis action IO and IW planning support, ranging from strategic-level planning through tactical execution. FIWC personnel are integrated into the staffs of numbered fleets, aircraft carrier battle groups, and amphibious-ready groups with their accompanying Marine Expeditionary Units. FIWC also provides the U.S. Navy's CERT capability. In addition to its headquarters in Virginia, the FIWC also maintains a detachment in San Diego, California to support operations in the Pacific. In July 2002, FIWC came under operational command of the Navy Network Warfare Command, a new staff that coordinates IO, C4I, and space for the sea service.

The Air Force Information Warfare Center (AFIWC) in San Antonio, Texas supports IO planning in the U.S. Air Force. The AFIWC is part of the Air Intelligence Agency, which is subordinate to the Air Combat Command (ACC). The AFIWC reorganized its deployed structure into IW flights, which are assigned to numbered air force, air expeditionary forces, and air component command headquarters. The AFIWC is uniquely positioned to provide IO and IW support due to its being co-located with the Air Intelligence Agency (AIA) and JIOC, mostly because the commander of the AIA also commands the JIOC. Until recently, the AFIWC provided the U.S. Air Force's CERT (AFCERT) capability. This has since changed as AFIWC has built a separate organization within the 67th IO wing that actually commands the AFCERT.

The JIOC is the key organization in the DoD specifically designed to support offensive and defensive IO planning. Located at Lackland AFB in San Antonio, Texas, the JIOC evolved from the Joint Command and Control Warfare Center (JC2WC). The JC2WC in turn was formerly known as the Joint Electronic Warfare Center (JEWC), which transitioned from purely EW to encompass C2W. In 2002, this organization was designated as the DoD "center of excellence for information operations." Originally falling under SPACECOM, until the shakeup of the UCP after September 11th, the JIOC now falls under STRATCOM to provide the Combatant Commander's and JTFs with teams of IO specialists. The JIOC also provides dedicated teams to each regional Combatant Commander and SoCom, and the remaining Combatant Commanders receive matrixed support. These JIOC teams thus provide technical and operational specialists to support IO planning, operations, and exercises.

An additional command that is essential for IO planning is the Joint Warfare Analysis Center (JWAC) located in Dahlgren, Virginia. Normally teamed with the joint information operations center (JIOC) these planners usually work in partnership with the Combatant Commander IO cell to assist the CJCS and commanders of unified commands in their preparation and analysis of joint operational plans. Specifically, the JWAC provides combatant commands, the Joint Staff, and other parties with information in order to carry out the national security and military strategies of the United States across the spectrum of operations. Falling under JFCOM, JWAC also provides direct support teams to assist unified commands with planning, and these teams usually work in conjunction with the Combatant Commander's JIOC team.

IO Planning Tools

Moving from IO planning organizations, this next section reviews a number of IO planning tools (IOPT) that are currently available to the military operator. Computerized planning tools to support IO are ubiquitous. Many of these are legacy systems from C2W and do not fully support full-spectrum IO planning. A number are also stand-alone systems, however, there have been attempts to develop software applications to support planning for all of the IO capabilities and related activities.

The Information Warfare Planning Capability (IWPC) began as an advanced concept technology demonstration (ACTD) at the United States Central Command in fiscal year 1997. Its original purpose was to demonstrate how IO planning, modeling, and analysis tools could aid in the effective execution of a Combatant Commander's battle objectives. These automated tools were to provide capabilities supporting the planning, development, synchronization, deconfliction, and management of an integrated IO campaign involving headquarters (HQ) CentCom's J-3 staff and the Combatant Commander components. The ACTD also was intended to support the modeling and analysis tools for the integrated air defense system (IADS) target recommendation development that is aligned with Combatant Commander IO taskings. Finally, the IOPT ACTD was to provide automated capabilities to enhance horizontal collaboration between multiple Combatant Commander components in planning and implementing Combatant Commander IO taskings.

The Information Warfare Planning Capability (IWPC) program will develop, field, and sustain an IO planning and decision support capability for the U.S. Air Force. The IWPC effort will include concept exploration as well as large-scale C4I system development, integration, and sustainment. This will include both systems architecture and planning tools and applications that must support all aspects of IO planning. The IWPC is planned to provide enhanced IO decision support capability, which must augment and interface to existing intelligence and operations systems. Currently under development by the Air Forces Electronic Systems Center (ESC), IWPC will provide a software suite that will enable collaborative IO planning, synchronization and targeting. The DOD IO Roadmap signed by the Secretary of Defense in November 2003 designated the IPWC as the future joint IO planning tool.

IWPC may eventually incorporate an IO planning tool developed at the JIOC, called the IO Navigator (ION). ION incorporates a planning process called the Joint Information Operations Attack Planning Process (JIOAPP). See Diagram 5-1. The ION software was specifically designed to support this process. The JIOAPP methodology used in the ION software is derived from the strategy-to-task planning methodology discussed later in this book and is an important component of IO planning. A defensive IO planning process is currently under development. ION is currently used by all JIOC teams deployed to Combatant Commanders around the world and it is also serving as the host software for the JFSC IO Planning Course taught semi-annually. ION has been purchased, or is under consideration for purchase, by several nations.

Strategy-to-Task Planning

The key to ION is that it incorporates the strategy-to-task methodology that was originally developed by the RAND Corporation (sometimes called objective-to-task) for the DoD,[5] as shown in Diagram 5-1. This type of planning is very different than what most staff officers are accustomed to. Instead of looking at what forces are in theater, in strategy-to-task methodology, the planner instead looks at the objective first, and asks the question, "What are you trying to accomplish?" In this context, IO and IW tasks may trace their origin all the way to the NSS. Thus IO and IW tasks are derived from a specific mission assigned to a regional Combatant Commander by the joint strategic capabilities plan (JSCP). To really understand IO planning, it's essential to understand this linkage. This is why so much time was spent by the authors explaining the national and strategic organizations that are involved in IO. High-level guidance, such as the NSS, is used to establish broad strategic objectives to protect the national security interests of the United States and describes the President's strategy for accomplishing these objectives.

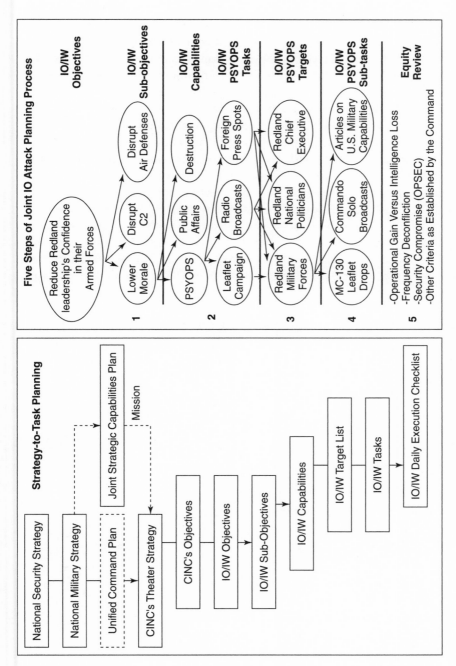

Diagram 5-1 Strategy-to-task planning process.

169

The national military strategy (NMS) is the DoD's plan for implementing the NSS, and flowing from the NMS are two key documents: the Unified Command Plan (UCP) and the joint strategic capabilities plan (JSCP). The UCP provides guidance to all unified combatant commands, establishes their missions, responsibilities, and force structure, delineates the general geographic area of responsibility for Combatant Commanders; and specifies responsibilities for functional commanders. Complementing the UCP, the JSCP provides guidance to the combatant commanders and the Joint Chiefs of Staff to accomplish tasks and missions based on current military capabilities. It apportions resources to combatant commanders based on military capabilities resulting from completed program and budget actions and intelligence assessments. The JSCP provides a coherent framework for capabilities-based military advice provided back to the President.

Each Combatant Commander develops a strategy explaining how he or she intends to accomplish the assigned missions. This is called a theater strategy for Combatant Commanders or a TSCP. The theater strategy

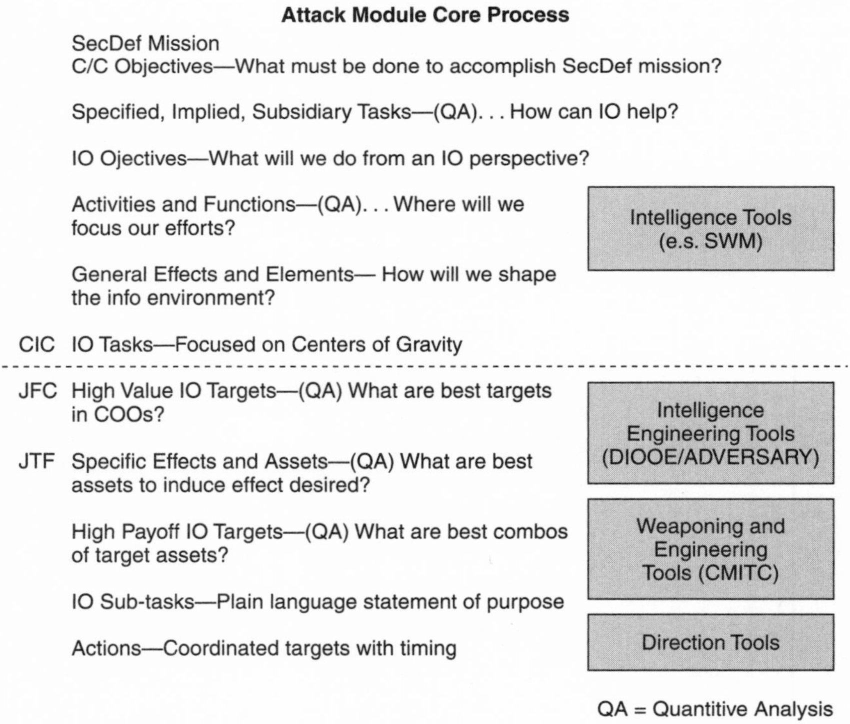

Attack Module Core Process

SecDef Mission
C/C Objectives—What must be done to accomplish SecDef mission?

Specified, Implied, Subsidiary Tasks—(QA). . . How can IO help?

IO Ojectives—What will we do from an IO perspective?

Activities and Functions—(QA). . . Where will we focus our efforts? | Intelligence Tools (e.s. SWM) |

General Effects and Elements— How will we shape the info environment?

CIC   IO Tasks—Focused on Centers of Gravity

- -

JFC   High Value IO Targets—(QA) What are best targets in COOs? | Intelligence Engineering Tools (DIOOE/ADVERSARY) |

JTF   Specific Effects and Assets—(QA) What are best assets to induce effect desired?

High Payoff IO Targets—(QA) What are best combos of target assets? | Weaponing and Engineering Tools (CMITC) |

IO Sub-tasks—Plain language statement of purpose

Actions—Coordinated targets with timing | Direction Tools |

QA = Quantitive Analysis

Diagram 5-2   Joint IO attack planning process.

enumerates the objectives that the Combatant Commander wants to accomplish in the execution of this strategy. For every objective the Combatant Commander develops, the staff planners will develop supporting IO objectives and sub-objectives as shown in Diagram 5-2. The planners then associate available IO capabilities and related activities with each IO sub-objective. Finally, each IO capability and related activity is assigned specific tasks to support the accomplishment of the IO sub-objectives. These tasks, after much coordination and orchestration by the staff, eventually appear on a daily IO task execution list, as well as on the synchronization matrix. Thus, the daily IO tasks, if properly planned and formulated, are directly linked to the NSS, via the strategy-to-task methodology. See Diagram 5-3 for an example of strategy-to-task methodology.

Strategy-to-Task Example

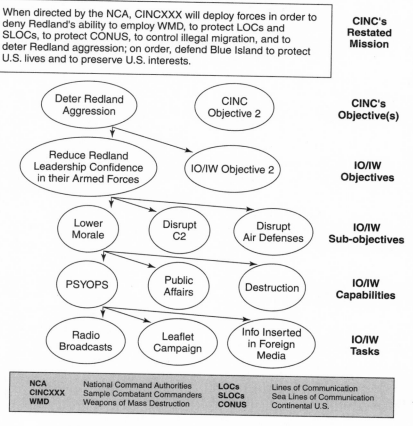

Diagram 5-3 Strategy-to-task example.

Tying Together Strategy-to-Task Planning and IO Planning Tools

With an understanding of the strategy-to-task process, it's easy to see the logic in the JIOAPP methodology used in the IO navigation (ION) software. The JIOAPP consists of five steps for planning offensive IO and IW. It facilitates planning at the unified command level and at subordinate components or joint task forces. The first phase is identifying the IO objectives and sub-objectives. This flows from the mission analysis process, during which the staff develops a restated mission and the Combatant Commanders objectives. As stated previously, the Combatant Commander's objectives answer "the what" that the Combatant Commander desires to accomplish. For the purpose of this book, we will use the IO sub-objective to "lower the morale" of the adversary for an illustrative example.

The next phase of the JIOAPP is generating IO tasks. These will be very broad tasks associated with the individual IO capabilities and related activities available to the planners. For example, lets look at only the PSYOPS tasks. Elements of a campaign may be to conduct a leaflet campaign, conduct PSYOPS radio broadcasts, and to place PSYOPS messages in the foreign press, which can be done by having a third party purchase column space in newspapers, magazines, and so on. The planners then identify IO targets associated with the individual IO tasks. Looking again at the PSYOPS tasks, our example identifies the Redland military forces, national politicians, and chief executive as PSYOPS targets. In the fourth step, the planners associate IO sub-tasks with each IO target. In our example, the PSYOPS campaign will target the Redland military forces with leaflet drops from MC-130 aircraft, *Commando Solo* radio broadcasts, and public information articles on U.S. military capabilities placed in selected foreign newspapers, magazines, and other media that one might expect members of the Redland military forces to have access to. Finally, the planners would conduct an equity review of the IO attack plan, which would include an analysis of the operational gain versus the potential intelligence loss, electromagnetic frequency deconfliction, an OpSec review of the plan, and other considerations as determined by the command. As shown in this example and in Diagram 5-3, the JIOAPP utilization in the ION software is an excellent case-in-point of how the strategy-to-task methodology is adaptable to computer-based IO planning tools.

IO and JOPES

The DoD conducts all joint planning within the framework of the Joint Operations Planning and Execution System (JOPES), which is a system of policies and procedures, combined with automated data processing systems, that is designed to provide joint commanders and planners with a method for planning and command and control of joint operations.[6] JOPES is well suited for strategy-to-task methodology and ideally supports IO and IW planning. There are three basic manuals in JOPES, and at the time of the writing of this text they were undergoing extensive revision. *JOPES Volume I* (CJCSM 3122.01) describes the DoD planning policies and procedures. It offers broad guidance and background information and is essential reading for all IO planners. *JOPES Volume II* (CJCSM 3122.03) establishes formats and offers guidance for preparing an operations plan (OPLAN). The final basic manual of JOPES is CJCSM 3122.02, *TPFDD Development and Deployment Execution.*

As mentioned, *JOPES Volume II* provides the basic formats for the various annexes and appendices in an OPLAN. The following section describes those portions that are important for IO planners. There has been much discussion in the IO community regarding the need for a separate IO annex to each OPLAN. Thus far, the general consensus is that because IO is a warfighting strategy, it should be incorporated throughout the various portions of the OPLAN, not lumped into a single annex. The wisdom behind this becomes clear when we look at just how many annexes in an OPLAN must address IO, because theoretically, each annex should address both the offensive and defensive aspects of IO.

Operations Plan, Time-Phased Force Deployment Data and the IO Cell

Planners generally view the development of time-phased force deployment data (TPFDD) as a logistics function. However, it is important for IO planners not to overlook this aspect of JOPES. The TPFDD is a database containing time-phased force data, non-unit related cargo and personnel data, and movement data for an OPLAN. It is important for IO planners because certain IO capabilities and related activities are specifically associated with units. As an example, most of the active component CA capability in the DoD resides within one unit, the 96th CA Battalion. Likewise, most of the active duty PSYOPS capability resides in the 4th Psychological Operations Group (POG) and both of these units are based at Fort Bragg, North Carolina. To a lesser extent, the same holds true for units providing EW support. IO planners will be interested in the arrival times of CA, PSYOPS, and EW units in theater, as their arrival will signal the IO planners as to when a particular IO capability or related activity will become available for employment. As a general rule, commanders will want CA, PSYOPS, and EW units available early in a deployment. IO planners may need to leverage the system to ensure that these units are included early in the flow of forces into a theater.

Finally, all of the planning staff's efforts will eventually result in an OPLAN. But the work isn't over yet. The most important part of any operation is the execution. From the planning that was conducted during the OPLAN development process, the IO cell will have a multitude of tasks that must be executed in a highly coordinated, synchronized, and time-phased sequence in order to support the accomplishment of the Combatant Commander's objectives. This is attained by producing a series of mission synchronization matrices through the course of the planning process. Initially, one matrix is produced depicting the major functions of the IO capabilities and related activities for each phase of the operation. Eventually, as more details develop, the IO planners will prepare separate synchronization matrices for each IO capability and related activity. Again, as more details are developed these matrices will be transposed into daily execution checklists for each IO capability and related activity. Daily execution checklists will be reviewed at each meeting of the IO cell and revised as necessary as the operation proceeds.

And planning becomes the life of the IO cell. Even as a plan is executed, the cell must constantly revise and adjust the IO particulars to ensure that the Combatant Commander's objectives are met. The high degree of detail required to successfully execute IO places the onus on every "IO warrior" to take full advantage of the information technology available to assist in planning. As new computer-based IO planning tools emerge, it is essential that these be integrated into the planning process. Additionally, published planning procedures and guidance are essential to ensure that the diverse IO planning effort conducted throughout the staff is coordinated and complementary. Just as sweat in training prevents blood in battle, sweat in planning will help prevent the unexpected during execution.

IO is considered a conceptual approach to military planning and operations, including their support functions, rather than a new area of military specialization. IO in current doctrine is the responsibility of operations staff assisted by other functional groups. Planning staffs also need to consider IO as part of future operations when conducting deliberate and contingency planning. At present within joint headquarters, a dedicated operations staff

Typical Joint IO Cell

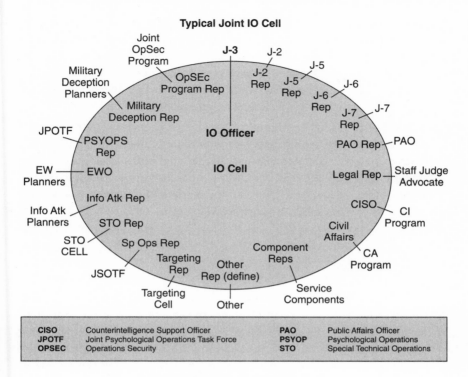

CISO	Counterintelligence Support Officer	PAO	Public Affairs Officer
JPOTF	Joint Psychological Operations Task Force	PSYOP	Psychological Operations
OPSEC	Operations Security	STO	Special Technical Operations

Diagram 5-4   Joint IO cell.

member is normally responsible for IO. At the tactical level, IO responsibility is met by command direction, awareness of IO, and representation on formation-level IO planning cells if required as shown in Diagram 5-4. The manner by which the IO planning function is met must be appropriate to the headquarters and ensure that IO is considered and integrated with the maneuver plan. The following staff responsibilities are generally considered as IO functions:

- *Operations:* Inclusion of IO in current operations plans with policy on IO, deconfliction and synchronization of IO including direction of IO-related assets on behalf of the commander and providing friendly force updates for the commander's situational awareness

- *Plans:* Inclusion of IO in future plans and contingencies and developing long- and short-term shaping strategies for the operations staff to implement.

- *Intelligence:* Through the joint intelligence preparation of the battle-space process provide intelligence on adversary critical vulnerabilities that may be exploited by IO including advice on the threat from IO and countermeasures as well as targeting intelligence on the specific charac-teristics of selected targets and technical control of intelligence collection operations and activity (including counterintelligence operations).

- *Communications:* Coordinating the electromagnetic spectrum man-agement and the management of friendly communications architec-ture with technical control over communications assets and their security, as well as technical control of information assurance mea-sures and advice on the vulnerability of friendly information systems, given input from intelligence processes.

- *Logistics:* Provide for IO-related assets and resources as well as advice on logistics infrastructure and plans, given input from intelligence processes.

IO planning needs to begin at the earliest stages of a potential conflict. At the strategic level, military IO planners need to engage relevant gov-ernment departments as early as possible to develop overarching shaping strategies. At the theater level, IO planning must be integral to the plan-ning process, which brings a range of activities that can be synchronized within the traditional maneuver of major force elements to achieve deci-sive points. The bottom line is that IO is not a separate planning activity, nor can IO be considered as a separate maneuver element, force element, or battle operating system.

IO as a Warfighting Strategy

Commanders will remain central to the operations planning process. Their guidance on IO will naturally include an assessment of how the commander views friendly and adversary vulnerabilities. This will translate into a prior-ity of targets and priorities of effort for scarce resources. The commander may also indicate the level of acceptable risk to be sustained and the per-ceptions to be manipulated. The commander's guidance may also include direction as to the deception target and deception objective and those key elements of information, personnel, and materiel to be kept secure. Central to the development of commander's guidance in relation to IO and targeting is the continuing balance between the implications of the Laws of Armed Conflict (LOAC) and the principles of war. Offensive IO will be conducted as a legitimate response, and hence, the selection of targets, the means of attack, the level of force applied, and the risk of collateral damage will all be in accord with national and international law. Development of protocols relating to emerging IO disciplines is expected to occur in the next few years.

It is the nature of IO that much of the effort to achieve related effects can be detracted from or confused by other IO-related activities. For example, experience shows thematic-based message related to military deception, public information, and PSYOP often conflict, even though they should not. Planners will ensure that thematic messages complement one another and that all IO activities work in harmony to achieve suitable outcomes in accordance with set priorities. Moreover, IO activities can often conflict with other efforts such as targeting, maneuver, and the desires of subordinate commanders. Again the planners will seek to plan for and avoid such conflict. Additionally, each IO activity must be synchronized with other efforts and wider operational efforts to achieve maximum effects. This process is fundamental to IO and is captured on a synchronization matrix or on a time- or event-sequenced chart. One of the most difficult features of this type of planning and staff-work will remain evaluating the effectiveness of IO measures. Increasingly, commanders will demand advice on the levels of risk of collateral damage and will need to be provided with a clear evaluation plan. The basis for evaluation lies in establishing and monitoring measures of effectiveness. It includes:

- maintaining an effective reporting system to identify degradation in information assurance,
- steerage of the efforts of the intelligence system,
- linkage to the combat assessment aspects of the targeting process, and
- integration with the security validation and reporting process.

During planning, the measures of effectiveness for each IO element must be defined. These are subsequently further developed and monitored as part of current operations. Measures of effectiveness are considered in an offensive context (for the targeting outcomes of offensive IO) and in a defensive context for the security aspects of defensive IO. These determinations ideally rely on such data as mathematical models, ongoing practical weapons testing, and historical analysis, all of which combine to enable staff to predict the effectiveness of IO activity. In PSYOPS this process may include "pre-testing" types of products to guarantee an appropriate effect.

In the future, when integrating IO into other operations, commanders will be offered an expanded range of options, which may include an ability to deceive, degrade, destroy, manipulate, or confuse an adversary's information and information systems. Hence, IO forms part of the wider operations process that feeds targeting related processes and activity. Though IO input to operations planning is recognized, the majority of the IO planners' daily work is related to current operations and hence such staff officers usually reside in the operations area of headquarters. Especially in cases where a task force is operating independently, an IO cell of specialists and involved parties is required to regularly meet to assist in the completion of IO staff responsibilities.

Legal Issues Connected with Information Operations

Throughout this book, the message is clear that the capabilities, opportunities, and threats involved with information operations are significantly different than have been the case in the past. It is also clear that, in many cases, the legal landscape is uncertain. Domestic laws within the United States and in other nations are changing in an effort to strike a balance between the needs of the law enforcement, national security, and business communities and the civil liberties of the populace. In many ways the situation is a hydra (the mythical monster who sprouted two new heads each time one was chopped off)—rather than settling issues, each inquiry into a legal matter raises more questions. Both as a matter of law and of policy there is uncertainty as to where IO fits. In each case, the question is asked whether what is being faced is a criminal act or a national security threat. It is not clear whether what is involved is a new form of mischief, a new example of the keen competition between businesses and nations, or at some point, an act of terrorism or an armed attack.

Over the past decade, the legal community has been paying increasing attention to the field of information operations. What used to be limited to discussions of computer crimes and more traditional forms of electronic warfare now looks at all of the new capabilities and techniques and attempts to divine the legal constructs which apply. In this attempt, two perspectives on the issues could be taken. First, what are the laws, policies, and rules of engagement affecting the potential use of IO concepts and tools in wartime, in operations other than war, and during peacetime operations? This is the perspective most often voiced by the client, the military commander and staff charged with execution of a discrete mission. Just as importantly in the long run, though, is the second perspective, namely that of the legal community and policymakers attempting to develop a comprehensive IO capability and strategy. From this perspective, the important question asks how the law must evolve to strengthen U.S. interests, policies, and capabilities with regard to IO.

An Overview of the Legal Landscape

It is beyond the scope of this text to cover the legal landscape in depth. Other sources exist which address these issues in great detail.[7] What we will do, however, is sketch the outlines of the legal issues in terms of the underlying principles which operational planners need to take into account in their planning—namely what must then be presented to the chain of command and their legal advisors so that these issues can be resolved to some degree before the planning concludes and execution of the mission begins. In this regard, the process now in place for developing the rules of engagement (ROE) for an operation, allowing the means by which operational planners can address the legal questions which may arise. Staff officers will encounter issues of both international and domestic law in planning any operation. What is important, then, for the planners is a basic understanding of the underlying principles and issues rather than the specifics of the legal analysis.

International law addresses the relationship between nation-states and, rather than a collection of rules, should be seen as a system by which the international community seeks a stability of expectations in its interactions. As with any other system of law, international law represents a struggle among the individual sovereign interests of the members of the community. The principle that each community member has certain sovereign rights and interests which can and should be advanced is balanced by the principle of reciprocity—that successful existence within a community requires members at times to subjugate individual interests so that the collective will can be advanced and through this coexistence an environment can be created which fosters individual sovereign interests.

The basic document which governs this reciprocal effort in the current international legal environment is the charter of the United Nations (U.N.). It was written to advance three interests:

- International peace and security
- International human rights
- Economic and social development[8]

Military operations are generally raised in the context of protecting international peace and security and it is from this context that the discussion will flow. The generally accepted view is that the U.N. charter establishes a balance between deterrence of aggression and promotion of defense. In this regard, Article 51 of the charter recognizes the inherent right of a nation-state, or a collective group of states, to protect its interests against armed aggression. The recognition of the inherent right to use force defensively is balanced by the prohibition contained in Article 2(4) against the aggressive use of force. This balance, recognized well before the charter was developed, can be accomplished unilaterally by individual states or groups of states or on behalf of the international community as a whole by the Security Council.[9] Established as the enforcement arm of the international community working through the United Nations, the Security Council has a great deal of authority.[10] Because of this, the Security Council can authorize action which would constitute a violation of Article 2(4) if undertaken unilaterally by a state or group of states.[11] If the Security Council does act in a given circumstance, all member nations are obliged to support that effort—there can be no neutrals in the case of a Security Council enforcement action.[12]

Peacetime Treaties Impacting IO

The previous analysis is important, not simply in the decision of whether or not to employ the use of force in an international context, but also in determining whether or not there are treaty obligations which should be taken into account in IO planning and execution. In this regard, there are two basic types of treaties involved—those such as the 1982 Law of the Sea Convention or the Outer Space Treaty, which address operating in a particular environment, and those such as the various international telecommunications conventions, which address the use of a capability. Both sets of treaties contain language that IO planners should be prepared to address. First, these treaties generally require that use of the environment or capability be reserved for "peaceful purposes." Secondly, each of these treaties contain some version of a requirement stating that a party operating within the environment or using the capability must do so in a manner which does not interfere with the legitimate use by another.[13] Clearly, military operations at any scale of intensity implicate these provisions. For many, the mere use by the military for the conduct of military operations seems to violate the treaty provisions. Similarly, the use of certain capabilities, jamming, for instance, are designed to deny others the use of a capability or environment. Under the aggression/defense analysis described earlier, the proper legal approach is to ask whether the operation being planned is being undertaken for a purpose or in a manner which is consistent with the principles of the U.N. charter. If so, then the action is generally going to be permissible, at least in concept.

This is not to say that this approach resolves all potential issues of treaty law. There may be other issues which arise from treaty obligations. Many of these treaties do, however, contain some sort of exemption for military communications, requiring compliance with the treaties "to the extent feasible."[14] Finally, it is generally accepted that treaties such as the various telecommunications pacts are intended to guide peacetime relations and would be suspended during a period of armed conflict.[15] Of course, as our military continues to be tasked with military operations other than war (MOOTW), it will not necessarily be clear whether the particular operation qualifies as an armed conflict. In this regard and based on the previous discussion, one critical question for planners at the operational-level staff to ask of the chain of command would be the extent to which treaties such as telecommunication conventions will be considered to apply in the context of the planned operation.

In addition, another treaty which will have to be taken into account in the planning stages of an operation will be any status of forces agreement (SOFA) or comparable diplomatic arrangement which may have been made concerning the basing of troops and operations in another nation. These agreements may limit the operations, which can be conducted from within the host nation or be launched from the host nation, and they may require coordination in the planning and execution of operations. In addition, these agreements will also affect the imposition of criminal justice on visiting forces. If civilian technicians will be a part of the IO cell, their status needs to be considered as well. Because of this fact, host-nation laws will have to be considered, much as U.S. domestic law is taken into account when operating from within the nation's borders. Again, these factors will vary from situation to situation and should be considered in the early stages of planning in order to avoid distraction at a later time.

Law of Armed Conflict

Ultimately, discussions on legal issues connected with the planning and execution of IO touch on the LOAC. At first blush, we are faced with a situation which will require the application of traditional LOAC principles to a new set of capabilities and threats. From the perspective of the operational planner, the issues need not be that different, again so long as the underlying principles remain the focus. All of the LOAC boils down to a societal balance between two objectives. First is the legitimate need to protect the ability of a sovereign nation, acting through its military commanders, to use force in order to successfully accomplish legitimate military objectives. The second is the equally legitimate need to protect the innocent from unnecessary suffering. All of the LOAC reflects the struggle to strike this balance in the context of differing societies, cultures, and levels of conflict. Each advance in warfare technology has resulted in an evolution of how these principles are applied.[16]

The principle of distinction requires a commander to be able, in choice of targets and of weapons, to be able to distinguish between combatants and other legitimate military objectives and civilian objects. The principle of necessity requires the commander to be able to demonstrate a definite military advantage in the contemplated action. The principle of proportionality requires the commander to consider the effect of the attack and the weapons used on the safety of civilians and other noncombatants.[17] However it is the unique nature of information as a power, and these interrelated principles of the LOAC, that suggest areas of concern in using IO capabilities or preparing to defend against new threats.

It is also important to note that the ground rules by which these principles are evaluated differ significantly. The United States considers legitimate military objectives to include combatants, defended places, and those objects which, by their nature, location, purpose, or use make an effective contribution to military action.[18] Clearly, military targets are simple in this regard; other infrastructure, including infrastructure relied upon by the civilian population, is more problematic. Many of our allies have adopted the protocols additional to the Geneva conventions. These set forth a more narrow definition for military objectives by adding the requirement that the total or partial destruction, capture, or neutralization of the objective must offer a definite military advantage.[19] Other language contained in the relevant sections of Protocol I clearly reflect an unwillingness to countenance the attack of infrastructure and other objects which, although arguably part of the war-sustaining effort, are not a clear part of the adversary's war-fighting capability. Although the United States takes the position that the definition contained in the additional protocols is reflective of customary international law, it also holds strongly to the position that infrastructure which indirectly but effectively supports and sustains the war-fighting effort may also be attacked.[20]

Much of the international community, then, sees a fairly distinct line between the legitimate targeting of war-fighting capabilities and the presumptively illegitimate targeting of infrastructure considered to be part of the war-sustaining capability. The United States takes the position that proper application of the balance between necessity and proportionality appropriately guarantees the safety of the innocent from unnecessary suffering—by seeking, to the extent possible under the circumstances, to inflict the least amount of collateral damage, so the argument goes, we maintain adequate flexibility for the military commander while protecting the innocent. Many others, though, see the issue as a slippery slope, and that is just for CIP attacks. It gets worse when you begin to consider CNA.

However, it could be argued that the very nature of IO capabilities could help the United States resolve this issue. After all, a discrete IO capability is sure to be more discriminating and less likely to result in unnecessary innocent suffering than high explosives. To date, however, that argument has not been persuasive. First, there is a lack of trust that these capabilities are in fact as discrete as the sales pitch suggests. We are often reminded that there exists a "law of unintended consequences," that is, you never truly know what the effect will be until you use it. In the current societal context, there is more comfort in reliance upon tried and true capabilities than in the use of a new capability without a track record.

Conversely, there is also the concern that even if the IO capabilities live up to their billing, they represent a capability which we, as the most prosperous and technologically advanced nation on earth, enjoy and which other poorer nations do not. In this context, there is concern that the use of our very discrete weapon against infrastructure relied upon by the civilian community will legitimize that infrastructure as a target even by adversaries who must rely on much less discrete weapons. These adversaries would argue that they were using the best they had and so ought not to be held responsible for any collateral damage which results. The solution for the international community is to ultimately tend toward a "lowest common denominator" approach to a greater degree. This unfortunately may, in the end, result in greater harm to the United States and its allies, if the adversary responds with a series of retaliatory strikes. One may even argue that the use of airliners by the al Qaeda terrorists to attack symbols of the United States on September 11, 2001, were examples of these principles in action.

Finally, there are also a number of other issues that the LOAC regulates. One of these warfare areas concerns the use of deception. An ancient facet of warfare, new computer-based plans or media operations have not changed the basic rules of deception.[21] It still remains unlawful to target the civilian population as such with any weapon, whether that weapon is a high explosive or a falsehood designed to cause confusion or civil unrest. In a similar vein, it is also unlawful to use a protected symbol or character as a means of deception. False identification as a medical or religious person, site, or other platform remains unlawful perfidy, regardless of the means of deception used.

Domestic Law

So far the discussion has been centered in the field of international law, identifying the context in which issues of the legitimacy of the use of force, of obligations under treaty law, and of obligations under the LOAC must be considered in IO planning. But what of the field of domestic law? To what degree do domestic legal issues affect IO planning? In this area, there are two basic questions we ask.[22] First, does domestic law adequately permit defense of IO capabilities? Secondly, does domestic law restrict offensive IO operations and action taken in self-defense? To answer to these questions, there are a number of major federal statutes that have been passed to protect national security, to protect the property of the private sector as well as the governments, and to protect the privacy and civil liberties of the citizenry.[23] These statutes both provide the basis for prosecuting criminals as well as the necessary checks and balances which our society feels are necessary to ensure against wrongdoing by the national security and law enforcement mechanisms.

In many ways, these checks and balances make the national security role more difficult. Those who are trained to act immediately in self-defense of their unit chafe at the requirement to use complex, cumbersome, and time-consuming processes such as those that exist in the criminal investigative arena. And yet, these checks and balances are a reminder that the democratic processes are by design inefficient, and that these checks exist to protect, however imperfectly, against governmental tyranny. Throughout our history of criminal justice, the struggle has been to design structures which allow identification and prosecution of wrongdoers without infringing on the rights of those who have not done wrong. Even more than in the past, the use of the electronic environment has made this goal key—unfortunately it is very difficult to achieve. First, we must recognize that the overwhelming majority of computer intrusions and attacks have been made for personal, rather than national, gain. Added to this, the fact that the criminal investigative processes, though relatively slow and filled with procedural checks and balances, are nonetheless reliable and accurate in the long run. From these ideas, we can then understand the current DoJ perspective, which is to treat all intrusions and attacks as criminal unless clear evidence exists of hostile state involvement. This point should be a reemphasis of the lessons learned in the organization section of this book.

Because this is the perspective taken, great effort is being made to ensure that processes are streamlined. Taking analogies from other areas of the law, from decisions made earlier in other contexts, the DoJ is working to ensure that the courts and legislatures strike the most effective balance between efficient identification, investigation, prosecution, and civil liberties. In his article "The Critical Challenges from International High-Tech and Computer-Related Crime at the Millennium," Michael A. Sussmann, senior attorney for the computer crime and property section of the U.S. Department of Justice, notes that the attorney general has identified four areas where progress is critical:

- First, the enactment of sufficient laws to appropriately criminalize computer and telecommunications abuses.

- Second, commitment of personnel and resources to combating high-tech and computer-related crime.

- Third, improvement in global abilities to locate and identify those who abuse information technologies.

- Fourth, development of an improved regime for collecting and sharing evidence of these crimes, so that those responsible can be brought to justice.[24]

In the planning of offensive IO missions, planners have been similarly obliged to keep in mind that these statutes have clear requirements for approval/coordination, without which the planned action is very likely illegal. Although the DoJ would in all likelihood be reluctant to prosecute a well-intentioned service member, the investigation necessary to determine the motive of the member and make this decision would represent a major distraction of resources and time. Not only the domestic laws of the United States, but also host-nation and target-nation laws represent potential sources of distraction at the least, and individual criminal liability at the worst. For example, Mr. Sussman theorizes that the situation could arise where a searching country took the view that a particular transborder search effort was permissible, whereas the searched country took the position that the execution of the electronic search is not only prohibited but constitutes unauthorized access to its computers and therefore is a criminal offense.[25]

The Solutions for the Operator: IO ROE Planning

It is clear from the foregoing discussion and from the discussions found elsewhere in this book that the legal landscape where IO is concerned is anything but certain. How, then, can military operators be expected to plan anything? That is where rules of engagement (ROE) and the IO planning process mentioned earlier in this chapter can help. ROE is defined as policies and procedures which govern the actions to be taken by U.S. forces during military operations. In general, ROE for peacetime operations or MOOTW are defensive in nature and address the decision-making process in determining the proper response to a hostile act or the demonstration of hostile intent.

It is important to keep in mind that the ROE are developed or modified by the SecDef and the CCs to fit the strategic and operational needs of particular events and operations. There is no desire for a "one size fits all" approach, but rather an iterative approach whereby the higher levels of the chain of command provide initial guidance and then respond to submissions from below to amplify, explain, modify, or substitute other ROE provisions based on the needs of each component or subordinate division. The ROE requirements of the naval component will vary from those of the Special Operations Forces component and each is expected to advocate for their required guidance. In this regard, over the last several years the operational planning process has spawned an ROE planning process which contemplates a cell—the "Knights at the Round Table"—gathering to discuss mission tasks and requirements in order to ensure that all operational views are explored and that appropriate ROE is requested. The approval process set forth in the ROE ensures higher headquarters' review and allows for interagency coordination prior to or as a component of any ROE approval. This process provides the CC with some measure of certainty in what response is permitted (or restricted) and under what circumstances.

This forum allows a process whereby the legal issues of IO can be identified and addressed. So long as the right questions are asked and incorporated into the ROE requests, a commander will be provided much of the guidance he or she needs to determine what legal issues are raised in IO planning. For example, during one exercise the following matrix was developed:

- Who/what is perpetrator/adversary? Criminal? Terrorist? State? Combination?

- Where is perpetrator located? United States? International waters/ airspace? Third country?

- What is the state of perpetrator?

- What is the impact on the United States? Minor disruptions or damage to national security?

- Who should respond? Law enforcement? Host country? Flag state? U.S. military?

- Is interagency coordination required?

Summary of IO Planning and Legal Concerns

From the foregoing, we have discerned that IO is a legal political-military tool and a means of warfare and that though existing law, regulations, and policies do not prohibit IO, the degree to which the law affects this warfare area will depend on the actual circumstances. Likewise the importance placed on planning for IO is crucial to the success of a campaign. It is incumbent upon the operator and planner to coordinate the various portions of IO methodology that were mentioned in this chapter. In addition, the importance of state custom and practice on the development of the law requires close involvement of legal advisors in development of IO capabilities and doctrine. Thus the ability of the IO cell to properly plan and execute using IO tools and methodologies will be symptomatic of the success rate for a campaign. To conclude, if nothing else is retained from this chapter, please remember that the IO cell and other staff planners must get early legal advice and integrate their operations in the master operations plan early on. That is the key to success.

CHAPTER 6

Implementing IO

Recent Campaigns

"China has realized from the outcome of the Gulf War several years ago that unlike the human wave tactics of the agricultural age and the iron and steel warfare of the industrial age, air raids and precision strikes from long distances are decisive factors in the outcome of wars. It also realizes that information warfare and electronic warfare are of key importance, while fighting on the ground can only serve to exploit the victory."[1]

Jen Jui-Wen, Chinese military leader

This chapter is an attempt to update the reader on events and activities that have occurred over the last several years concerning IO, on real-world operations and how IO has been used during this period. Specifically we will cover updates to Russian doctrine, IO aspects of the NATO coalition in Operation Noble Anvil in Kosovo, Chinese writings on IO, and most recently the Australia Defence Forces' campaigns in Bouganville and East Timor. This chapter will analyze these operations, focusing on the use/misuse or action/inaction of IO in each of these geographic areas.

The Growing Role of Information in Russia

The operations of the Russian military offer many fascinating examples of how important information has become in modern combat, especially with the glaring eye of global television ever present. Some of these changes are reflected in Russia's current national security documents that reflect more of an increased concern over information security issues than earlier versions. The most recent (October 1999) military doctrine draft states that the exacerbation of the information opposition/confrontation is an important feature of today's international context, a destabilizing factor used to achieve destructive military-political goals and affect current operations and the overall security environment. The draft includes information-technological (attacks on computers, nets, infrastructure, etc.) and information-psychological aspects of the external threat to

Russia, and states that the greatest internal threat is actions to disrupt or disorganize the Russian Federation's information infrastructure.

The military-strategic features of the new draft doctrine focus on the features of modern war, namely indirect strategic operations, means of IW, and the development of a massive information preparation (information blockades, expansion, aggression) operation. Confusing public opinion of certain states and the world community (a.k.a. IPI) and achieving superiority in the information sphere in either wartime or during the initial period of war are also other important missions. This new draft elevates information security up to a basic military mission and ensures that information support remains a constant priority in the realm of information-economic principles.

The concept of national security that was approved by the Russian Security Council in October 1999 also addressed the country's information security and technology needs. The chapters titled "Russia's National Interests," "Threats to the Russian Federation's National Security," and "Ensuring the Russian Federation's National Security," include the following information-specific interests:

- Observing the constitutional rights and freedoms of citizens to obtain and use information.

- Developing modern telecommunication technologies.

- Protecting the state information resource against unauthorized access to political, economic, science and technology, and military information.

- Preventing the use of information for manipulating the mass consciousness of society.

- Attempts by a number of countries to dominate in the world information space and to crowd Russia out of the foreign and domestic information market.

- The development of "information warfare" concepts by a number of states envisaging the creation of means of exerting a dangerous effect on the information spheres of other world countries, means of destroying the normal functioning of information and telecommunications systems, and means for the safekeeping of information resources or of gaining unauthorized access to them.

- Implementing citizens' constitutional rights and freedoms for information activities.

- Improving and protecting the domestic information infrastructure and integrating Russia into the world information domain.

- Countering the threat of the initiation of opposition in the information sphere.[2]

For Russian security specialists examining the national security environment on the verge of the new millennium, no issue is more important or more fraught with uncertainty than that of the current and future information environment. There are several good reasons why this is so. First, the free-flowing, border-crossing exchange of information has offered people and organizations in the former Soviet Union unstructured access to fresh information never before available. This relatively unfettered environment permits citizens and decision-makers alike access to a wide variety of ideological, political, religious, and other information that was once forbidden by strict internal and external barriers. Second, the Russians now perceive that information itself has developed into a very important type of national or strategic resource. Compare this to the ideas about the power of information that were discussed in the first chapter. This is because information can potentially increase the precision and effectiveness of both traditional (missiles and rockets) and nontraditional (nonlethal and psychological) types of munitions which could thus upset parity in strategic arms. Third and finally, many Russians believe that a single global "information space" is emerging. If they are correct, then a country can exploit this space to alter the global balance of power.

Information Superiority

The Russians also believe that countries that possess information superiority may be more inclined than ever to employ military force. This is so because military objectives may now seem more attainable without significant loss of life and with no apparent risk. This is what, for many Russians, provides an explanation for the recent NATO intervention in Kosovo. The Russians have also realized that few legal restraints exist that can regulate the use of an information attack. Russians are convinced that this actually encourages the growth of concepts such as cyberterrorism, which includes the use of terrorism against information processing systems. Finally, many Russians also understand that they are far behind in the global race for information superiority and are beginning to appreciate and fear the potential consequences. These three final reasons have prompted the recent Russian attempts at the United Nations, as mentioned earlier, to limit the development of information operations procedures.

It is because of these considerations that the subject of IW/IO has become almost as significant and important to Russian military planners as the issue of nuclear proliferation. Russian theorists warn decision-makers not to submit to external forms of coercive information diplomacy. At the same time, subcommittees of the State Duma are commissioning studies on both IW and psychotronic warfare (which is similar to

PSYOPS), and the Kremlin advisors and the security community are studying how information security issues may affect the country's political, technical, economic, and military policies. Some members of the Russian academic community are also engaged in studying the potential impact of information operations. The analyst E. A. Belaev, a member of the Russian State Technical Commission (under the President of the Russian Federation), believes that the *informatizatsiia* (the informationization) of society has led to the collection, processing, maintaining, and exchange of information between actors—people, organizations, and governments—in the single information space. As Belaev defines them, the most critical information technologies within this space are those that support:

- Governmental and military command and control organs
- The financial credit and banking structure
- Command and control systems of various types of transport, energy, and ecologically dangerous industries (nuclear, chemical, biological, and others)
- Warning systems for emergency situations and natural disasters

Any underestimation of the information security of these systems, Belaev argues, could lead to unpredictable political, economic, ecological, and material consequences, and perhaps even turmoil. This analysis sounds very similar what the U.S. doctrine espoused in PDD-63, *Critical Infrastructure Protection*. It shows that nations today must protect their national information resources as strategic resources. In addition, the burgeoning access to global information networks such as the Internet have only underscored the necessity for protecting information resources from manipulation, corruption, deception, or even outright theft. Furthermore, the Internet has also become an arena for potential conflict, especially over modern information concepts and unauthorized access to databases—witness the recent Solar Sunrise and Moonlight Maze incidents (discussed earlier).[3]

Information Space
The conflict in Kosovo has done little to assuage Russian concerns about the significant role information will play in national security issues in the twenty-first century. In the case of Kosovo, for the first time, the United States and NATO justified military activities by different geo-strategic principles other than simply national interests. In fact, writing in *Foreign Affairs,* Joseph Nye asked if it is possible to define interests convention-

ally in the information age, especially in light of humanitarian concerns that, due to the impact of the mass media, divert public attention away from real strategic issues. He summed up his views, stating:

> The Canadian media guru Marshall McLuhan once prophesied that communications technologies would turn the world into a global village. Instead of a single cosmopolitan community, however, they may have produced a conglomerate of global villages, each with all the parochial prejudices that the word implies, but with a greater awareness of global inequality . . . all in the presence of television cameras and the Internet.[4]

Nye noted that the United States now has an interest in the use of outer space and cyberspace, expressed in language similar to the language once used by the British to express its desire for freedom of the seas. Notably, both are the channels through which words and ideas pass and democratic principles can be promoted. However, the same medium of cyberspace has also been used to promote the advancement of democratic interests, such as humanitarian affairs to the level of a state interest at a startling pace. The Clinton administration clearly appeared to agree with this assessment, as is clear from their justification for the use of force in Kosovo. In summary, Nye added, "a democratic definition of the national interest does not accept the distinction between a morality-based and an interest-based foreign policy."[5] It is thus clear that new geo-political principles are beginning to emerge in response to the influence of information.

This concept of information space has two dangers; first it can be used to monitor the state's information resources, thus becoming a conduit for information espionage; and second, the information interaction can destroy or disorganize the information resources of elements of state structures. These effects can be realized in peacetime, especially if critical application systems are affected, thereby distorting or destroying information used for state management or decision-making. IO protection is what many theorists contend is the greatest promise for this new warfare area and also its weakest link. Just as the Russian federation is attempting to develop information security, the current inability of the USG to effectively capitalize on this capability will ultimately determine the true effectiveness of IO. Information space has no recognized boundaries, no institutions to protect state interests such as border or customs checks. The nation is transparent to information resources and analysts believe that one day, states may try to regulate the movement of information flows. This is because there are currently three ways that information impacts national security. The first is the security of vital state information resources and information systems, counters to which are being actively developed by countries all over the world. Second is the predominance of the information approach as the emerging primary scientific method of solving national security problems.[6] Finally, information can impact on a state or person's social awareness by manipulation of reality or fact, which in turn can have a deep impact on a state's national security decision-makers.

Russian IW Terminology and Theory

Many Russian theorists differ in opinion about the elements that comprise IW. Following are two of the different variants, both of which are the products or thinking of either theorists or practitioners who could be considered Russian information warriors. The first theory is from the former first deputy minister of defense and national security chief, Andrei Kokoshin, who was ultimately responsible for research and development of these information systems. He divided information warfare into the following five subcategories:

• Electronic warfare
• Intelligence
• Communications
• Operational command and control systems
• Facilities for the protection of command and control systems against enemy influence[7]

The second theory comes from the civilian Russian minister of defense, V. I. Tsymbal, who wrote of additional categories. Information warfare, in his view, must be considered as an integrated whole of systems working together that includes the following eight subcategories:

• Intelligence and counterintelligence gathering
• Maskirovka (Occupation) and disinformation
• The use of EW systems
• The debilitation of communications and scrambling of enemy data
• A determination of to which state a military objective belongs
• The destruction of an enemy's navigational support
• The use of psychological pressure on the enemy
• The destruction of enemy computer nets and software programs[8]

However whatever the number of groupings a theorist might believe in, just as in the United States, until IO has been extensively tested in combat, there will be continual doctrinal development.

General Major N. A. Kostin, chairman of the Radio-Electronic Department, General Staff Academy has written a general theory of IW. He listed both *informatsionnoy bor'boy* and *protivoborstvom* as simply the same definition offered to the United Nations, "a form of struggle between sides that involves the use of special methods and means for impacting the information medium of the opposing side and protecting one's own side in order to achieve the assigned tasks."[9] Therefore his goal was to provide information security for one's own side and lower the information security posture of the opposing side. If you compare that to the current U.S. military doctrinal for IO espoused in JP 3-13, you will notice many similarities. Kostin notes that the battle over information is now so important that the battle for ore, oil, and markets could fade in comparison. General Kostin added that the information struggle is a special category of war because it is an independent type of war, a component element of any other form of war, and it is waged constantly in peacetime and wartime. Once again, compare this to the U.S. doctrine.

Kostin also believes that political factors have the greatest impact on the substance of IW, driving their goals, tasks, and issues. This idea is comparable to the development of PDD-68 by the NSC and State Department. Therefore it is the political factors that determine the means, methods, and characteristics of conducting the battle and its scope and duration, and also that provide the necessary material support and financial resources. On the other hand, economic factors determine the scientific and technical development of the computerization of society and the state. Kostin has described information factors as influencing both political and economic readiness by determining the scope of the struggle, the procedure and methods of its conduct, and the capabilities for utilizing these factors when influencing the enemy's information environment.

The logical elements, according to Kostin's general IW theory, that form the foundation are categories, laws, patterns, and principles. Categories objectively reflect the essence and core characteristics of the most important manifestations of IW. They also represent a body of military-theoretical thought that includes general terms such as information and IW, and in particular terms such as protecting information and attacking information. Categories can reflect the structure, substance, and requirements of IW. The laws of the materialistic dialectic also present themselves as well, according to Kostin, as objective laws and patterns of military activity valid for IW. These include:

- The law of the defining role that politics plays in IW
- The laws on the course and outcome of war and IW

These laws depend on the economic, socio-political, scientific-technical, and military capabilities. Recognizing the patterns that are inherent in IW is where the current primary efforts have been conducted. Therefore the effectiveness of IW is determined by the proportionality among the goals, tasks, and systems used, including means available, which takes into account the enemy's countermeasures.

Informational-Psychological Components of IO

Probably the most interesting and neglected, when compared to Westerners, Russian element is the information-psychological component of IO. The Russian military excels in the study of these areas compared to Western theorists. To date, the United States has not conducted extensive analysis in this area of IW except for those personnel involved in psychological operations. On the other hand, the Russian military scientists have been studying not only the ability of information warfare to affect the values, emotions, and beliefs of target audiences (traditional psychological warfare theory), but also methods to affect the objective reasoning process of soldiers. That is, Russia is interested in ascertaining how to affect not only the data-processing capability of hardware and software but also the data-processing capability of the human mind.

Three books published in the Russian federation recently serve as an example of this fixation on the mind. Endorsed by the State Duma's Security Committee, the first book was titled, appropriately enough, *Informatsionnaya voina* (Information War).[10] This book examined how to manipulate the mind by toying with the algorithms (including how to model them) that define human behavior. Humans, the author notes, like computers can have a "virus" inserted in their information system (reasoning process) if the proper algorithms of mental logic can be affected. The authors dubbed this human information virus a "psycho virus," which according to mathematical formulas could perhaps be inserted as a "suggestive influence" to alter the mind's algorithms or prevent objective reasoning. The second book, entitled *Psikhotronnoe oruzhie i bezopasnost' rossii* (Psychotronic Weapons and the Security of Russia)[11] bore the endorsement of the State Duma's Information Security Committee. It was co-authored by Major Vladimar Lopatin, the Chief of the Information Security, which is a subsection of the Security Committee of the Duma.[12] Lopatin and his co-author V. D. Tsigankov define psychotronics as an inter-disciplinary area of scientific knowledge, which, when mediated by consciousness and perceptual processes, investigates distant (non-contiguous) interactions among living organisms and the environment. Another book, published in 1999, covering information-psychological problems was titled the *Secret Weapons of Information Warfare.* Focused squarely on the psychological impact on the mind by information issues, the chapters of the book are as follows:

- "Basic Directions in the Development of IW under Modern Conditions"
- "Understanding Phenomenology in Man and Controlling His Behavior"
- "Education on the Use of Psycho-Physical Weapons"
- "Methods for the Precise Orientation of Covert Effects on the Human Psyche"
- "Psychotronic Means of Subconscious Effects on the Human Psyche"
- "The Integral Method of Psycho-Physical Weapons"

These two books plus another book, *Psychotronic Weapons and the Security of Russia,* also written by Lopatin and Tsigankov, are part of a series known as the "Informationization of Russia on the Threshold of the twenty-first century." What is important here is that these three books underscore the Russian belief that informational and psychological matters should be of concern to civilian and military alike as valid subjects for close scrutiny and that their effects, both positive and negative can and should be experienced both in peacetime and wartime. Colonel Igor Panarin of The Federal Agency of Governmental Communication and Information (a.k.a special service) (FAPSI), speaking at a conference in 1997, stated that there is a need in Russia to develop information-psychological subunits in government and military directorates. The role of these departments would be to develop strategic and operational measures to prevent or neutralize attempts to control the psyche of Russian society (what he termed the "strategy of psychological defense"). If formed, this main directorate in support of psychological security could ensure the psychological component of Russian national security. All of these efforts by the Russians are understandable when you look at how they define an information weapon. They view it as a specially selected piece of information capable of causing changes in the processes of systems (physical, biological, social, informational, etc.) according to the intent of the entity using the weapon. Information weapons are aimed not only at hardware and software systems as listed below, but also at "wetware" or the mind.

In addition, methods of persuasion are considered another Russian IO tool. The primary information weapon in this regard is a concept known as reflexive control, which is also called "intellectual IW." Reflexive control is defined as a means of conveying to a partner or an opponent specially prepared information to incline that person to voluntarily make the predetermined decision desired by the initiator of the action. There are scientific and mathematical components as well as varied military and technical uses. Russian academics have often noted that the goals of reflexive control are to distract, overload, paralyze, exhaust, deceive, divide, pacify, deter, provoke, suggest, or pressure an opponent with information. Other less known but reported information-psychological related activities include:

- Military unit 10003, which studies the occult and mysticism, reportedly to understand the recruiting and "brain washing" techniques of these groups.

- Anti-ESP training in the strategic rocket forces, designed to enable missile launchers to establish mental firewalls in case someone from the outside attempts to take over their thoughts.

- Astrologers in the Ministry of Defense, who predict ambushes, plane crashes, and other phenomena.

- Practice with the "25th frame effect," which tries to insert a subliminal message into every twenty-fifth frame of a movie or computer-generated scene.

- Applying electromagnetic impulses to the head of a soldier to adjust his or her psychophysical data.

- Remote viewing and psychotronics.

Russian military researchers have focused on the informational and psychological stability of individuals and society as a whole for a variety of cogent reasons, but the primary reason is the psychological security of Russian citizens. This is due to the striking change that has occurred in the country's dominant ideology, a change that did not occur in the West. Understandably, therefore, the absence of a similar ideological shock in the United States and Europe has prompted less attention to this subject. However, that may change in the West as the trend is for the proliferation and use of computer games, which can influence the youth who are increasingly interested in the subject. As a special note, more American researchers are now pondering the influence of information technology on the minds of its citizens, a phenomenon accelerated by the sort of youth violence that has taken place more frequently in the past few years.

Military-Technical Components of IO

According to Marshall Igor Sergeyev, Russia's minister of defense, the war in Kosovo demonstrated that a new phase of the revolution in military affairs is upon us. The United States, he notes, is in the midst of a significant military-technical breakaway in the sphere of information support of combat operations that must be countered in the future. Sergeyev's comments in a December 1999 issue of the military newspaper *Red Star,* devoted to military-technical issues on the eve of the twenty-first century, also discussed the main domestic and foreign threats to Russia, and the primary missions and problems of Russia's military-technical policy.[13] Sergeyev used the term "information" fourteen times in his discussion of military-technical issues, and his emphasis is not surprising. Over the last few years, Russian specialists have studied and written about information issues profusely.

Sergeyev also notes that the NATO campaign in Kosovo signified the beginning of "contactless" or virtual information-technical warfare. The biggest advantage the coalition possessed came from information-support systems, such as reconnaissance platforms, which contributed mightily to the overall success in this operation. Unable to compete at the present time, Sergeyev believes Russia must look to asymmetric options. This recognition is important because the theorists now believe that Russia can not support both the military-strategic and military-technical parity with the leading military powers of the West on a "symmetrical" basis, especially in the area of non-nuclear armaments. Therefore, some theorists believe that it may be necessary to search for a reasonable combination of evolutionary and "revolutionary" paths for more effective asymmetrical directions for the development of weapons and military technology, as well as for technologically outfitting the Russian armed forces.

The emphasis noted by Sergeyev should be on reconnaissance and command and control systems, with the latter specifically at the operational-tactical and tactical levels. The goal is to create an integrated information environment and a single system of military standards to transmit data. Other requirements would be to universally equip a force that was information-oriented and that could essentially make use of miniaturized computer systems. Sergeyev advocates the need to closely integrate information systems and nuclear weapons, stating that information-technical developments of both support and defensive systems would help to guarantee the effective use of nuclear weapons and could be a "new aspect of nuclear deterrence." Weapons based on these new physical principles signify a qualitative leap in the forms and means of armed conflict and would definitely change the parameters of "parity." Sergeyev also notes that Russia's main priority in the field of prospective weapons should be concentrated on guided and electromagnetic energy weapons, cyberweapons, and stealth unmanned combat platforms. His final concluding remark is that a new phase of the revolution in military affairs has begun and Russia must not lose time. Time frames are of such importance that any further delays in starting a full-scale modernization of the armed forces could lead to a fatal and insurmountable disadvantage to the Russian military forces.

Systemological Aspects

Russian scientists recognizing the increased importance of information systems have also adjusted their emphasis on the growing influence of information on the military-technical aspect of military doctrine. Therefore they have focused more attention on the interaction of combat systems instead of on the old reliance on simple force and force ratios. According to this logic, warfare is now viewed as the interaction among the military systems rather than forces in a confrontation. This idea has been extended to the modeling and simulation conducted at the General Staff Academy, where Red versus Blue (i.e., the US vs the USSR) is no longer the only war game played. Today, high-tech systems are also modeled against other high-tech systems.[14] To place this emphasis in context, within Russian military systemology information is viewed as the "nourishment" that gives life to all elements of the system. In particular, this applies to reconnaissance, command and control, and support and strike systems. To put this in perspective, IW can be viewed as a system, which, according to this view includes three components:

- Information support of the functioning of one's own combat systems
- Information counteraction against the functioning of the enemy's combat systems
- Information protection or defense of one's own combat systems against the informational counteraction of a possible enemy[15]

Therefore under modern conditions, the skillful use of one's information potential and information resources, including information means and systems, will increase the force combat potential many times and will increase the effectiveness of using weapons, combat equipment, and combat systems as a whole. This definition and its implications are similar to how the United States views the potential of IO. At the same time, the vulnerability of command and control systems with respect to deliberate and random activity in the information sphere continues to increase. Therefore, the Russians just like Americans, now understand that it is necessary to protect or defend one's information potential—to protect it everywhere and continually—not only in peacetime and wartime, but also from a probable enemy and also against unexpected changes in the current situation including social, economic, and diplomatic conditions, as well as from a lack of skill and/or professionalism on the part of subordinates and chiefs.[16] Of course not all Russian forces have accepted IO as the future of warfare. Though there is a growing interest in military systemology, not only in modeling information warfare but in national security in general, there are still some in the military forces who look at it as not much more than witchcraft.

What has been approved recently is, however, a very definite change in the Russian military and federal government doctrine and policy for the future of warfare. The writings that have been promulgated over the last few years indicate a shift from the previous emphasis on technological developments to a huge interest in what IO can do for their forces. Especially with the extremely tight fiscal situation that the Russian military finds itself in today, it is no wonder that asymmetric warfare and IO have assumed a much larger role in their recent doctrinal publications. Although the Russian military forces have tended to be written off in the last few years, one cannot discount their capabilities.

As mentioned earlier, the Russians have not organized their forces like the United States and they have also focused on different aspects of IO. The emphasis on psychological aspects is entirely different from where other military forces have decided to orient their attention. This is a whole new area of study and the Russian federation is known for employing many highly trained academics, including mathematicians and scientists, who have now concentrated their attention on investigating all aspects of IO. Just because the Russians do not organize their operations along the same lines as allied nations does not necessarily mean that they are wrong.

IO in Kosovo

If the Russian doctrinal debate shows how information and its use in war is evolving in this new revolutionary period, then an alternate example could be the use or misuse of IO in the planning for the Kosovo campaign. This was a massive air campaign conducted by a coalition of United States and NATO air forces against the former Yugoslavia over its policies of genocide in the Serbian province of Kosovo. The coalition flew over 38,000 combat sorties in a seventy-eight-day period of bombing, inflicting massive destruction on Serbia's economic infrastructure.[17] Rather than bringing stability to the region, as IO doctrine dictates, NATO's operation actually created greater regional instability and the potential for future conflicts. What caused this problem? It was due, at least in part, to the fact that no concerted peacetime IO campaign was implemented to deter

conflict with Serbia.[18] Once the conflict began, however, IW was successfully executed to help bring the conflict to a peaceful conclusion. The
problem here is that IW tends to rely heavily on physical destruction supported by other IO capabilities and related activities. Failing to execute a
concerted peacetime IO campaign against Serbia, the United States and
NATO were unable to avoid inflicting severe damage that ultimately
makes a post-conflict period much more difficult to manage both politically and economically.

The Use/Misuse of IO in Operation Noble Anvil

The strategic bombing campaigns first described by the renowned Italian
air-power theorist General Giulio Douhet and executed by the Allies
against Germany in World War II are a thing of the past for the United
States.[19] Douhet envisioned a total warfare where a nation's military,
industry, and population were attacked to bring about a swift and total
defeat. IO doctrine, on the other hand, does not advocate attrition bombing attacks and wholesale destruction against an adversary. Indeed, the
advent of precision-guided munitions and effects-based targeting has
added a whole new dimension to using physical destruction as an information weapon. The mere ability to destroy one of an adversary's high-
value targets while leaving the surrounding area virtually unscathed sends
a very potent psychological message. First, it demonstrates the precision,
lethality, and superiority of American weapons technology. More important, from an IO perspective, limiting collateral damage and physical
destruction gives the adversary less ammunition for hostile propaganda
directed against the United States. Second, the U.S. military has now so
conditioned the international media to low-collateral damage and precision engagement that when the occasional accident occurs and a nonmil-
itary target is hit, the media will tend to amplify the effects of the accident.
By their sheer excellence, recent U.S. aerial campaigns have inadvertently
set an inescapable standard for minimizing collateral damage. However,
there is much more to IO than just a targeting or destruction campaign.

Therefore, both the domestic and foreign publics expect the United
States to avoid inflicting massive collateral damage and civilian casualties
because it has the technological means to do so. Failure to accomplish this
strategy makes the United States a target of criticism by domestic and foreign media and politicians alike. The very manner in which the United
States uses physical destruction may in fact provide an information tool
for an adversary. When the United States uses physical destruction to
manipulate the behavior of an adversary, it must defend itself against the

hostile propaganda of that adversary and strive to maintain absolute credibility. Therefore, it is critical that the public affairs and PSYOPS messages describing the use of physical destruction is absolutely accurate. Early in the war, NATO press releases boasted about the effectiveness of its air war achievements in Kosovo. These later proved fairly inaccurate as the Serb combat vehicles withdrawing from Kosovo after the conflict were counted by NATO observers. *Aviation Week and Space Technology* in July 1999 reported that NATO had dropped 3,000 precision-guided munitions, had hit 500 decoys, but had only destroyed 50 Serb tanks.[20] NATO had been fooled by Serb denial and deception operations and the Serbs knew it. In what may have been an overzealous desire to demonstrate positive results from a two-month-old air campaign that was beginning to draw considerable international criticism, NATO put its credibility on the line with statements the Serbian military knew to be inaccurate. Given that the National Army force in Kosovo was the target of U.S. IPI and PSYOPS efforts, any loss of credibility with the target audience ultimately harmed these operations.

In addition to its PSYOPS efforts, the United States also uses public affairs to inform an adversary of its military's destructive capabilities and to dissuade them from behavior that is contrary to national security goals. A good example of this is the large amount of media coverage that was given to the arrival of the Apache attack helicopters in Macedonia during the Kosovo conflict with Serbia. The United States intended the presence of the lethal Apaches to have a psychological impact on the Serbs. The Apaches received great amounts of media coverage due to their awesome destructive capabilities. However, this backfired when one crashed during a training mission, generating a rash of bad press regarding the poor state of training of the Apache aircrews.[21] Whether or not this was true, it somewhat negated the intended psychological impact.

Although the original thought was that this would be a short strategic bombing operation, in reality, the war quickly began to drag on, as the effects of the strikes did not phase the Serbians. In fact, it was not until almost eight weeks into this campaign that strategies were developed to try to use new methods to bring pressure on Milosevic himself. As one analyst noted, the bombing didn't bring about the desired results, and so other tactics were used against the dictator. Specifically, attempts were made to conduct an information campaign, one that would discredit his policies, while at the same time undermining his economic means to continue the conduct of the war. High-level diplomats conducted near simultaneous press briefings emphasizing the fact Serbia as a nation was condoning Milosevic's genocidal actions. In the meantime, NATO aircraft

conducted bombing missions against specific factories and industries that
were funding the upper leadership. NATO sent the same Serbian government officials detailed messages tailored to the specific recipients, in an
attempt to influence them to shift away from their allegiance to Milosevic.
All of these actions taken together, along with the military, diplomatic,
and economic pressure, are what many people believe helped to bring an
end to this conflict. One may not know for sure, because much of the
details are still classified, but reports are starting to leak out after over two
years, that the information campaign against Milosevic was successful in
a number of different areas.[22]

To conclude, Kosovo will probably rank as the "second information
war." Through the use of advanced information dissemination including
faxes, e-mail, and web pages, as well as perception management campaigns, this conflict was fought for the hearts and minds of a worldwide
audience. Yet where the ultimate changes were actually made was in
the detailed, tailored targeting of the key individuals that could affect the
decision-makers. That is what was different about this operation and the
use of information. It was, for the first time, a war in which information
was recognized as the primary weapon used to bring about a decisive end
to the conflict.

An IO After-Action Report for Kosovo

The bottom line is that an overall information strategy was never attempted
against Yugoslavia, despite almost seven years of warning. As emphasized
throughout this book, IO is a long-term strategy that must be put into
motion during peacetime. The failure of the United States to implement an
information plan was largely attributable to the lack of political direction.
To be effective, there must be national-level direction to ensure that the
required interagency coordination takes place early in the planning process.
Another major problem was the lack of coordinated United States and
NATO strategies. In addition, the absence of a United Nations resolution
signaled a lack of international support, and the failure to recognize the significance of Russian participation also caused unnecessary turmoil for the
coalition. Since there was no clear link between the military and political
strategies of United States and NATO, impatience with the diplomatic
process inhibited the execution of an IO strategy.

At the operational level, serious blunders were made that precluded
the long-range planning of an IO campaign. The ruling out of ground
forces violated the principles of operation security and deception. There
was also an absence of a contingency plan to address the public relations
aspects of military failures, for example, the Chinese embassy bombing.

The confusion that followed that incident greatly damaged NATO's credibility. Furthermore, the public information campaign never connected the Danube bridges' destruction to protecting the Hungarian minority in Vojvodina. In addition, NATO did not produce a military "video" comparable to the Yugoslav capability, which quickly led to their credibility being questioned when "ground truth" failed to match the press releases (i.e., where are the destroyed tanks?). The absence of the NATO Secretary General in the beginning of the campaign in the United States media also created a perception of lack of unity in the coalition. These and other mistakes hindered the possibility conducting a viable IO campaign against Milosevic and his forces.

How an IO Campaign May Have Succeeded

To better understand the concept of IO as a warfighting strategy, let's take an example of how an information plan might have been employed from the start in Kosovo. Imagine that NATO forces in Kosovo as part of the regional stabilization strategy are engaged in helping restore normalcy to the lives of the Kosovar citizens. Many of the NATO activities would consist of coordinating projects to restore economic infrastructure, restore public works, and reestablish the local governments. The pre-conflict demographics indicate that the population of Kosovo was approximately 90 percent ethnic Albanian and 10 percent ethnic Serbs. NATO would expect the ethnic Serbs to distrust NATO intentions, as the Serbs were the primary target of NATO hostilities during the short conflict. It would therefore be imperative for NATO to appear even-handed in assisting the different ethnic groups. This would mean that the Serbs should enjoy the same benefits and security that the NATO operation affords the ethnic Albanians.

Now imagine that part of the plan would require a NATO mechanized infantry brigade to establish its command post in a town having an ethnic Serb majority and a mostly Serb local government. How might NATO employ information operations to help stabilize this area, reduce friction, and deter hostile actions towards the NATO force? First, NATO would employ its intelligence system to conduct an intelligence preparation of the area, which could be called an intelligence preparation of the battlespace. This would occur through a variety of means, including ground and aerial surveillance and reconnaissance, signals intelligence, human intelligence, and open-source intelligence. PSYOPS studies, if available, would contribute valuable information on the social, religious, and ethnic characteristics of the area. NATO Civil Military Cooperation (CIMIC) personnel would coordinate with any nongovernmental, private volunteer and international organizations present in the area before the conflict to help gather information about the area. After developing a preliminary intelligence estimate and conducting PSYOPS or CIMIC assessments, NATO would begin planning for the operation in earnest. The intelligence community would also continue to develop intelligence on the area.

NATO would then begin an information preparation of battlespace as mentioned earlier. This would help prepare the area inhabitants and particularly the ethnic Serbs for the arrival of NATO forces. Planners could choose from a number of different means to convey this information. NATO public information personnel would prepare media releases to inform the populace of the details of the employment, including answering when, why, and how many NATO forces would be introduced into the area. NATO might also disseminate the media releases to the local Albanian and Serb media, drop PSYOPS leaflets, conduct radio and television broadcasts, distribute NATO-produced newspapers in the Albanian and Serbo-Croatian languages, and use any other means available to get the desired information to the people. The public information releases would focus on informing the populace, and the PSYOPS releases would focus on very specific facts and themes aimed at influencing the Serbs to accept the NATO presence. These PSYOPS themes might stress the economic or security advantages of having coalition forces in the area and would encourage Serb cooperation.

With the information preparation of the battlespace completed, NATO could begin introducing forces into the area. The initial force would have a security element and would probably be accompanied by CIMIC and PSYOPS troops, who would establish contact with the local Albanian and Serb leadership in addition to any NGOs in the area. The aim would be to help assist in the introduction of additional troops and to further assess the geographic area. NATO would continue a gradual introduction of forces to avoid giving the impression that an occupying force was entering the area. As the mechanized infantry brigade headquarters became operational, CIMIC troops would identify projects that NATO could coordinate to help gain the acceptance of the Serb population. Within the limits of force protection requirements, the NATO mechanized infantry brigade personnel would conduct activities to increase the direct contact between NATO soldiers and the local populace, which usually tends to help gain acceptance. The brigade commander would periodically conduct personal meetings with the local Albanian and Serb civilian and military leadership and any other influential members of the community, in order to clarify NATO's intentions, help resolve any issues or misunderstandings, and make personal assessments of the local attitudes.

The NATO public information and PSYOPS troops would also furnish information to inform and influence the local populace to cooperate with NATO. This information would be disseminated by a variety of means, which might include public information releases to any local Albanian or Serb media and public information broadcasts in the Albanian and Serbo-Croatian languages over radio KFOR. In addition, public information broadcasts over local civilian radio with purchased broadcast time, radio and television broadcasts either commercially or from *Commando Solo* aircraft, handbills, and any other media available to disseminate information to the people. While all of these activities were underway, NATO would employ operations security to deny potential adversaries any critical information that would indicate the plans and activities of the NATO forces moving into the area. Deception might be employed to mask the arrival of a NATO force into the area and to deny potential adversaries the ability to accurately assess the size and strength of the NATO force. This sort of approach to informing the local populace would continue as long as the mechanized infantry brigade remained in the town.

This scenario obviously did not occur, but it shows how an integrated IO campaign may have been successfully implemented in Kosovo. Ultimately it was the lack of an overall information plan in this political war that probably led to the relatively unsuccessful conduct of this campaign. One would think that the U.S. military would have learned more in the last few years about how to conduct a proper IO campaign, but that was not the case in Kosovo during 1999. This is especially egregious considering the advanced state of IO doctrine in the United States. However, as many people realize, just because a country publishes doctrine, it does not necessarily mean that they understand or employ it. The United States is a case in point. Although we may not be studying our own doctrine, that is certainly not the situation for IO in general. What is especially interesting is the huge amount of interest in IO doctrine by other nations, especially China, who are studying this new warfare area at a feverish pace.

Information Warfare and the People's Republic of China (PRC)

Given the Chinese military leadership's continued fascination with the U.S. conduction of the Gulf War, there is little doubt that in the next conflict, the PRC is likely to employ its own version of the information-based warfighting techniques. Just how strongly the PRC's military establishment has been persuaded about the lethality of the information warfare techniques was described by Lt. General Huai Guomo in his book on information war. Describing a series of techniques that could be used by the PRC in a future conflict, Guomo relates:

> Before a battle begins (sometimes dozens of hours in advance) and proceeds, commanders will first use offensive information-war means (precision guided weapons, electronic jamming, electromagnetic pulse weapons, and computer viruses) to attack enemy information systems, affecting or destroying their decision-making mechanisms and procedures, thus forcing an end to the fighting in line with the aspirations and terms of the offensive sides. And meanwhile, to protect their own information and information systems from enemy destruction, they will set up in combat space among all targets and weapons real-time detectors—links among shooters. Such offensive-defensive information warfare will become the focus of coming wars. The struggle for information supremacy will gradually become the crux of the battle, in a sense as strategic deterrent.[23]

Consider how that quote compares to the discussion from Chapter 1, in which it is noted that the U.S. tacticians have also stated that information superiority is the key to success in future conflicts. The other significant influence on the thinking of the Chinese military analysts is their conclusion that the People's War under modern conditions has undergone an irreversible change. Much of that analysis came from observing the U.S.-led coalition effort in the Gulf War. Soldiers equipped with low technology, like the soldiers of Iraq and the PRC, will encounter a decisive tactical disadvantage when faced with high-tech-equipped American and European forces. Consequently, it has been argued in the latest PRC doctrine that new technology is particularly important, especially in local wars.

The Chinese military establishment is also very preoccupied about emerging as a high-technology-based force in the twenty-first century. A recent examination of the writings of its military analysts underscores the fact that they are avid readers of American professional military journals and the futurists, much of whose work has deeply influenced the thinking of senior U.S. military leaders. One scholar, Su Enze, believes that the military revolution has already happened. "Guided and represented by information warfare," he writes, "a military revolution is also taking place in military ideology, military theory, military establishment, combat pattern and other military fields on a global scale."[24]

The PRC scholars are also quite sensitive to the notion of information as a prime strategic source in warfare and the importance of intelligence in contemporary warfare. One author writes:

> In strengthening the information concept as a multiplier of commanders, we must take information as a multiplier of combat effectiveness and see

it as a strategic resource more important than men, materials, and finances, so that it can be properly gathered, employed in planning, and utilized. We must make efforts to raise our capacity to obtain, transmit, utilize, and obstruct warfare information and must include these elements in the whole process of command training.[25]

Chinese defense specialists, like their American counterparts, are looking for the "perfect weapon" in information warfare. One hears the echo of Admiral William Owens's advocacy for "a system of systems." Cai Renzhao advocates that the PRC "should try to gain insight into the development situations of foreign military forces, to try to understand future warfare by accurately recognizing the differences between ourselves and foreign military forces to fully bring our own superiority into play and explore the 'perfect weapon' on a digitalized battlefield."[26] He recommends that the PRC follow the European Union's example in a "focused way," and learn lessons from the United States and Europe in developing information-related research. According to Cai Renzhao, the PRC should "fully bring into play the guiding role of information warfare research in building the military . . . to seek measures by which to launch vital strikes in future warfare, so as to damage the enemy's intelligence gathering and transmission abilities and weaken the enemy's information warfare capacity."[27]

Chinese IO as a Warfighting Network
Similar to what is discussed in the first chapter, conventional organizations in the Information Age are undergoing major changes and the notion of hierarchy has become outmoded. In its place are emerging multi-organizational networks, with the U.S. military performing a trail-blazing task of undergoing radical changes in response to the radically divergent techniques of warfighting in the Information Age. Called "joint warfighting," this term serves as an umbrella phrase under which a multitude of changes are taking place. Under the auspices of the Goldwater-Nichols Act of 1986, the U.S. military has not only been busy converting the task of joint warfighting into an operational form, but is also continuing to do more with less. This means that different organizational and functional agencies have been formed to serve as a credible warfighting force. The fact that JV 2010 and its follow-up doctrine JV 2020 have emerged in an abbreviated discussion as the premier white papers of the U.S. military and the importance of IO within that doctrine has not been lost on the Chinese military.

These defense analysts also appear to understand that information warfare is at the cutting edge of military doctrine and all the implications that this means for traditional institutions like the military. Xu Chuangjie writes, "The revolution in information technology has increasingly changed with each passing day the battleground structure, operational modes, and concepts of time and space while dealing blows to the traditional 'centralized' and 'tier-by-tier' command structure."[28] He cites the U.S. Army's building of a "ground force operational command system," as an attempt "to organize various command control systems of the . . . ground forces into an integrated mutually linked network to realize 'shared information' from the national command authorities on top down to a grass-roots unit."[29] He emphasizes the significance of strengthening, completing, and perfecting the building of a command and control system for the PRC. He also recommends that the command and control system "at and above the battalion level of various services and service arms" be turned into an integrated, mutually linked network. In addition, the traditional vertical and tiered command system must be converted into a network command structure in order to meet the demands of time and flexibility in command, and finally the centralized type command system should gradually be developed into a dispersed type command.[30] Once again, this ties into the horizontal integration concept mentioned in the first chapter of this book.

The Chinese military establishment has also been quite conscious of its country's vulnerability to potential acts of sabotage during peacetime, as well as attacks during a military conflict, and is taking steps to reduce this vulnerability. Wei Jincheng writes, "An information war is inexpensive, as the enemy country can receive a paralyzing blow through the Internet and the party on the receiving end will not be able to tell whether it is a child's prank or an attack from an enemy."[31] Discussing the use of viruses in a netwar or even a cyberwar, another defense scholar writes, "Computer viruses can be used to track down enemy's target system and the enemy's guided missiles may end up attacking the side which has launched them or deviate far from the intended target. . . . After locating its target, a virus may replicate rapidly, erasing the normal operating database, thus overwhelming and crippling the computer system."[32] The same article discusses the variety of measures taken by the U.S. military in reducing its vulnerability to potential attacks, including sabotage attempts from terrorists and hackers.[33]

It is obvious that the military establishment in the PRC is watching the recent research and development of "virus warfare" in the West with rapt attention. The main focus of their interest, once again, is the U.S. military. This section concludes that a number of suggested preventive measures are needed against a future netwar or cyberwar. First, it advocates raising the consciousness of military computer security throughout the Chinese armed forces. Second, it asks the PRC military establishment to pay special attention to removing "hidden perils to hardware and software security," by creating security filters and careful tests on all imported electronic equipment. Finally, it recommends the initiation of "special-topic research on computer viruses."[34] What is extremely interesting from a Western viewpoint is how these Chinese articles tend to mirror similar concerns of U.S. analysts.

The Future of IO in China

As one ponders the future dynamics of United States-PRC relations in the context of information warfare, at least three observations come to mind. First, even though the Chinese military establishment wishes to emerge as a high-tech-based warfighting machine, based upon current intelligence, it will likely take time for this to occur. Second, this reality should not let anyone forget that the current commitment of the Chinese armed forces to high-tech-based warfare, if continued with the same zeal, will likely pose a serious challenge for the United States in the coming years. The state of readiness of the Chinese armed forces in the realm of information warfare at the present time may be at a very primitive level compared to the U.S. armed forces, but they are writing doctrine and experimenting with IO on a constant basis. A leading U.S. information warfare specialist, Martin Libiki, postulates that:

> Militaries, especially those of widely different nations, cannot prosper by copying each other. . . . Their endowments, circumstances, and strategies differ greatly. Each must adapt the general to the specific. We know the Chinese can copy our thoughts, but whether they can innovate in pursuit of their own objectives is not yet obvious.[35]

Third, China's smaller neighbors must not only watch the Chinese military preparedness closely, especially in the realm of information warfare, but also try not to remain too far behind in this field. This is not to suggest that the PRC and its neighbors are likely to fight one or more wars in the near or distant future. Rather, it is to suggest that the military establishments of a number of countries of that region are in the process of being equipped with state-of-the-art weaponry and they are well-served to emulate U.S. military preparedness in the realm of information warfare-related technologies as much as possible.

The Chinese have a very long history of adapting different technologies for their own use. They view information warfare as a tool to counter the overwhelming military superiority of the United States. If they can influence world opinion through international public information, PSY-OPS, electronic warfare, and so on, then the PRC will try to shape the environment to facilitate their goals. IO is all about perceptions management, as discussed in a number of areas in this book, and the side that can best influence the adversary decision-maker will be in the most advantageous position to ultimately affect the final outcome.

Introduction to IO in Australian Defence Force

One nation that will be affected by China and its growing capability within the IO arena is Australia. In this section, this nation's development of its own doctrine and force structure will be discussed. What is very interesting about Australia is that it is a nation very large in land mass but with a small population and even smaller military forces. Therefore, IO would seem to be a great tool for it to use to maximize its effectiveness in the region and around the world.

In common with the armed forces of many other states, one of the most important operational concepts adopted by the Australian Defence Force (ADF) in the last half decade has been IO. In particular, over the last several years, IO has been incorporated into Australian Defence Organization (ADO) policy and into joint/single service doctrine as well. Instruction on IO has been introduced into the curricula of a variety of courses taught at generalist and specialist single-service schools, as well as in the joint training establishments. IO doctrine and training has been tested on exercises and most importantly, it was recently used during real-world operations. In short, the ability of the ADF to conduct IO in concert with other operational activities has become a fundamental part of the ADF's approach to warfighting.

This section is intended to outline the ADF's approach to IO and recent Australian experience in the conduct of IO in operations. The intent is to focus on developments within the ADF at primarily the operational and tactical levels. This is not to say that there has not been significant policy activity at the strategic level. A case-in-point here would be the significant activity outside of the ADO regarding the protection of the Australian National Information Infrastructure (NII).[36] However, it is at the operational and tactical levels that developments with respect to IO in Australia are perhaps the most readily apparent.

The Evolution of IO and Related Concepts in Australia

Tracing the development of any form of military doctrine is inevitably a difficult task. Descriptions of the adoption of particular concepts or approaches will tend to focus on official announcements encapsulated in policy documents. In practice, this approach reflects the end or outcomes of doctrinal evolution and innovation rather than the realities of the doctrine development process. It ignores or marginalizes the significance of less formal influences on doctrine development, such as the observations and experiences of exchange officers and the modification or outright copying and adoption of overseas ideas. It also tends to disregard the influence of individuals from outside the military establishment, be their influence direct or second-hand via the pressure that they may exert on service personnel.[37] With those caveats in mind, the following section provides a broad overview of the development of IO doctrine within Australia.

Like most other defense forces across the globe, serious thinking about the implications of an information-based revolution in military affairs for the conduct of operations commenced as the result of observations of the coalition's performance in the Gulf War.[38] Key observations include the increasing military value of attacking or manipulating an adversary's information and information systems, the need to deny an adversary the ability to do the same, and the requirement to integrate such activities with other military operations.[39] A further key factor worthy of note was the rise of the so-called "CNN effect," the pervasiveness of global electronic media and the influence that it exerts on public opinion, thereby shaping political and (therefore) military decision-making.[40]

The conclusions drawn from the Gulf War were reinforced by Australia's operational experiences in Somalia and Rwanda as well as by observation of overseas experiences in Haiti, Bosnia, and elsewhere. Experience and observation of peace operations demonstrated the utility of influencing the information environment at all levels of conflict, not just the middle- to high-end of the conflict spectrum. In particular, it was noted that a technologically inferior adversary might still have the ability to influence the information environment in their own favor by exploiting the vulnerabilities and weaknesses of high-technology systems.[41]

The Australian response to these observations was gradual. Through the mid-1990s, a variety of capabilities that had languished since the end of the Vietnam War were resurrected, in particular PSYOPS and to a lesser extent CA.[42] Between 1995 and 1997, there was extensive discussion of a variety of information-related operational concepts, including terms such as C2W and IW, in professional military journals and conferences organized by the ADF. Through the later half of 1995, Australian Defence Headquarters (ADHQ) undertook a major study of the implications of the C2W concept for the ADF. In 1996, the Defence Science and Technology Organisation (DSTO) commenced Project Takari: a long-term project to enhance the ADF's information capabilities for the battlespace of the future.[43] But it was not until late 1997 that information-related concepts were clearly articulated in an official defense policy statement.

At that time, the Australian government released the document *Australia's Strategic Policy* (ASP97), which identified the achievement of a "knowledge edge" (KE) as Australia's highest priority in defense policy.[44] ASP97 described a KE as being the product of three elements of capability—intelligence, C4 systems, and surveillance/reconnaissance capabilities.[45] As has been noted by the then Air Vice Marshal Nicholson, this provides a rather limited account of the role that information-related activities can play in the context of armed conflict.[46] However, this account of the KE provided a basis for further conceptual development, in that it left open the possibility of other means by which a KE might be achieved.

In early 1998, the commander of the newly formed Headquarters Australian Theatre (HQAST) released the document *Decisive Maneuver*, which for the first time articulated an Australian approach to warfighting at the operational level of war.[47] The capstone concept of *Decisive Maneuver* is composed of five core concepts and three supporting concepts. Underpinning these core and supporting concepts is *decision superiority*, which is defined as the ability to make and implement more informed and accurate decisions at a rate much faster than an adversary.[48] This is very similar to the information superiority concept espoused by the United States. In addition, shades of the observe, orient, decide and act (OODA) loop are seen in these Australian doctrinal writings. In turn, decision superiority is enabled by four broad sets of activities:

- Information management
- Intelligence
- Protective measures for C4 systems and processes
- Offensive C2W directed at adversary C4 systems and processes

Hence, whilst the decisive maneuver concept precedes the adoption of IO by the ADF per se, it goes well beyond the KE concept articulated in ASP97 as shown in Diagram 6-1. This is because it includes measures to *shape* the information environment, as well as the collection, processing, management, and dissemination of information. Significantly, it also functionally divides information-related activities into three broad categories:

- Offensive
- Defensive
- Supporting activities

As will be seen later, this division of effort is also reflected in the IO construct adopted by the ADF.

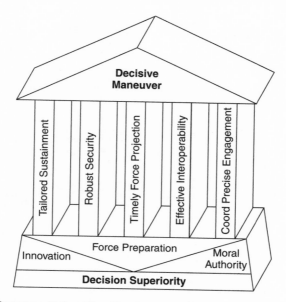

Diagram 6-1   The Decisive Maneuver Concept.

The formal doctrinal adoption of IO by the ADF occurred during 1998. An initial draft version of the *Information Operations Staff Planning Manual* (IOSPM) was circulated throughout the ADF by the Australian Defence Force Warfare Centre (ADFWC).[49] This became the doctrinal basis for the planning and conduct of IO across the ADF at the time. The approach adopted by the ADF divided IO into three distinct but interdependent components: offensive IO, defensive IO and IO support. These three components are further divided into specific IO capabilities and activities, which are shown in the Diagram 6-2. Not unsurprisingly, the model of IO adopted by the ADF fairly closely resembled the one presented in JP 3-13. In addition, at the departmental level IO was finally given official recognition in the document *Defence: Our Priorities*, which noted that the development of IO capabilities was a key priority for defense.[50] Hence by the beginning of 1999, IO had formally become part of the ADF's approach to warfighting.

Now that IO has been formally adopted as part of the Australian approach to warfighting, there still remains the issue of linking IO to other warfighting concepts. In a presentation given in early 1999, a model was demonstrated to connect IO with other information-based warfighting concepts at the Defence Communications Development Seminar held in Canberra.[51] The model noted that though a combination of IO and

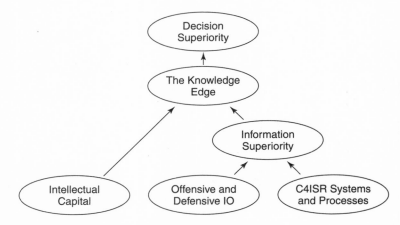

Diagram 6-2   Decision Superiority and Subordinate Concepts.

Command, Control, Communications, Computers, Intelligence, Surveillance, and Reconnaissance (C4ISR) systems might provide information superiority (IS), this in itself did not guarantee decision superiority (DS).[52] The KE which led to DS was seen to be a product of both IS and superior intellectual capital within the ADO.

This approach to DS implies the possibility of a warfighting concept which could be referred to as knowledge operations (KO) or knowledge warfare (KW) as shown in Diagram 6-2. Such a concept would include not only the activities which are currently encompassed within IO, but would also incorporate other measures centered around the attack and defense of the organizational intellectual capital that transforms IS into a KE, and thus to DS.[53] These ideas extend well beyond the concept of IO as it is presently understood within the ADF and elsewhere.[54] In particular, it implies a far closer functional relationship between the operational conduct of IO-related activities and the development and maintenance of capability, in particular the human dimension of capability than is presently the case. The full development of a coherent KO/KW concept that can be readily employed in an operational setting is probably some time off. Nevertheless, the conceptual possibility of KO/KW highlights the dynamic nature of operational concepts relating to the information domain.[55]

Having incorporated IO into doctrine and policy, the next step for the ADF to take was to acquire additional IO capabilities. Some indications of the future capabilities that the ADF might acquire were given in late 2000 with the release of the 2000 Defence White Paper and the associated Defence Capability Plan.[56] The white paper identified, for the first time,

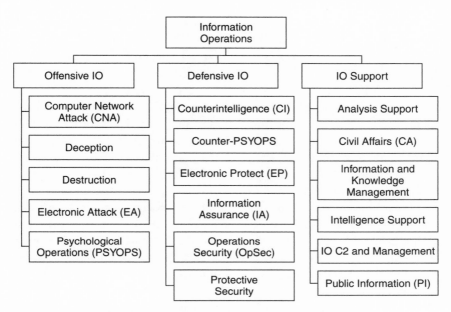

Diagram 6-3   Unclassified ADF IO Doctrine.

information capabilities (including IO-related capabilities) as a discrete capability area for the ADF.[57] The Defence Capability Plan expanded upon the white paper, and provided further detail (at the unclassified level) of the future IO capabilities that would be acquired by the ADF over the next decade. Some of the key IO-related projects are highlighted in Diagram 6-3. Thus, by 2001, IO had clearly become a fundamental part of the ADF's approach to warfighting, as manifest in policy, doctrine, and capability development and acquisition.

The Australian Doctrinal Approach to IO

Having examined the evolution of IO within the ADF, attention will now be focused on the details of the current doctrinal approach to IO as accepted by the ADF. It should be noted that, unlike in United States joint doctrine, there is no separate definition offered in Australian IO doctrine for IW, as the activities and effects encompassed within IO are held to be applicable across all levels of the conflict spectrum. Additionally, the term C2W has fallen into disuse within the ADF. Whereas previously C2W was at times treated as a distinct operational concept, it is now (at least, informally) regarded as an application of IO at the operational and tactical levels against a particular target set (i.e., the adversary's C2 system).

Current ADF IO doctrine (still in draft) has matured somewhat since the circulation of the IOSPM in late 1998.[58] ADF IO doctrine still resolves IO into the three distinct but interdependent components of offensive IO, defensive IO and IO support. However, offensive IO and defensive IO are no longer resolved into specific capabilities or activities. Rather, it is recognized that most IO capabilities and activities can be employed for both offensive and defensive purposes. Hence, the offensive or defensive nature of IO relates to the ends to which it is directed, rather than being an inherent characteristic of a particular capability or activity. The capabilities and activities associated with IO support have remained essentially the same.

It should not be inferred from the present doctrinal model of IO that all the capabilities listed are currently possessed by or are intended for future procurement by either the ADF or the ADO as a whole. However, what the model does acknowledge is that the increased integration of high-technology computers with communications equipment into the everyday business practices of the ADO represents a source of potential vulnerabilities that can and will be exploited by adversaries. Accordingly, there is a need to understand how adversaries might exploit these vulner-abilities, and to incorporate the means that an adversary might employ against the ADO into the Australian doctrinal account of IO.

The range of capabilities or activities incorporated in the Australian model of IO is essentially similar to that which is offered in the JP 3-13, although there are some differences. Australian IO doctrine does not include any direct equivalent of the U.S. term "special information opera-tions" (SIO), though Australian doctrine does recognize that some IO capabilities or activities are sensitive, and will require special authorization for their employment. In addition, the number and range of activities listed under IO support in Australian doctrine is somewhat more extensive than the two (CA and public information) listed under IO-related activities in JP 3-13. In the Australian model, these two activities are incorporated within the IO support area.

IO has also been incorporated into single-service capstone doctrinal documents for the Royal Australian Navy (RAN), the Australian Army, and the Royal Australian Air Force (RAAF).[59] At the present time, the present draft ADF IO doctrine is in the process of being revised, with the intention of publishing it as an Australian defence doctrine publication (ADDP) in the near future.[60] The details of the model of IO that are finally articulated in future doctrine may vary somewhat from that which is presently espoused. However, it is likely that the fundamentals will remain essentially the same.

The Australian Experience of IO—Two Case Studies

As mentioned earlier, Australia is included in this chapter because not only is it a smaller force that is attempting to add IO doctrine within its services, the ADF has also used IO in two recent operations over the last three years. Though there are obviously many portions of these missions that remain classified, we will, over the next few pages, discuss the operational IO issues pertinent to these two task forces.

Bougainville—Background

The first case study on the operational employment of Australian IO doctrine is Australian involvement in the Truth Monitoring Group (TMG) and leadership of the Peace Monitoring Group (PMG) to Bougainville, codenamed Operation BELISI. A map showing Bougainville relative to the rest of Papua New Guinea is shown in Diagram 6-4.

Whilst the roots of the Bougainville conflict precede Australia's granting of independence to Papua New Guinea (PNG) in 1975, the immediate origins of the conflict are dated in the late 1980s.[61] In late 1988, the long standing grievances of the Bougainvillean population against both the PNG central government and the Australian-owned copper mine at Panguna erupted into guerrilla warfare. The conflict had commenced with demands by Bougainvillean locals for compensation for the environmental

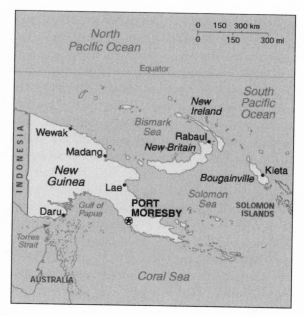

Diagram 6-4 Location of Bougainville within Papua New Guinea.

damage caused by the mine and for a greater share of its profits. However, by 1990 these demands had transformed into a demand for the secession of Bougainville from PNG.

For about the next six years, the conflict waged on and off at varying degrees of intensity. On one side was pitted the PNG security forces (the PNG Defence Force—PNGDF, and the Royal Papua New Guinea Constabulary—RPNGC) and their auxiliary arm, referred to as the Resistance. On the other side were the Bougainville Revolutionary Army (BRA) and its political wing, the Bougainville Interim Government (BIG), both of which, in turn, were highly factionalized. Further complicating this conflict environment were assorted criminal elements collectively known as *raskols* who preyed on the Bougainvillean population, sometimes under the banner of either the Resistance or the BRA. In the midst of all these contending parties was the Bougainvillean population, who of all the parties, suffered the worst from the effects of Bougainville's descent into anarchy.

The consequences of the Bougainville conflict were harsh, through both the direct effects of the conflict and the indirect effects of the destruction of the health and social infrastructure on Bougainville. It has been estimated that up to 20,000 people perished during the conflict and up to 50,000 were displaced either within Bougainville proper and Papua New Guinea or the nearby nation of the Solomon Islands. Given an estimated pre-conflict population for Bougainville of between 160,000 and 200,000, the conflict inevitably touched the lives of all Bougainvilleans.

Between 1990 and 1996, there were a number of efforts made to broker a peace settlement on Bougainville. Such efforts at restoring peace to Bougainville resulted from efforts both within PNG and from the international community, in particular from the Southwest Pacific (SWP) region. In the latter category, the most notable effort in this regard was the Australian led Operation LAGOON, which supported the Arawa Peace Conference held in October 1994.[62] The Arawa Peace Conference failed as the result of the non-attendance of senior BRA/BIG leaders, which was in part a consequence of their distrust of Australia's intentions and avowed neutrality in the Bougainville conflict. Significantly, the creation of an environment of trust and neutrality would later become a key plank of the TMG/PMG information campaign in Bougainville.

PNG government frustration with the performance of PNG security forces on Bougainville led to consideration of alternative means to end the Bougainville conflict. In turn, this led to what is generally referred to as the Sandline Affair.[63] In late 1996, PNG Prime Minister Chan secretly contacted the British military consultancy firm Sandline International, with the intention of securing support for the PNGDF in an assault on the Panguna mine, which he assessed would break the back of the BRA. Key PNGDF leaders balked at this proposal and in March 1997 launched Operation Rausim Kwik ("Get Them Out Quick") which led to the detention and ejection from PNG of Sandline International personnel. The outcomes of the Sandline Affair ultimately led to the collapse of the Chan government. But at the same time, it paved the way for an invigorated Bougainville peace process, assisted by diplomatic efforts undertaken by New Zealand.

In July 1997, the various PMG and Bougainvillean, less the hard-line BRA faction of Francis Ona, met in New Zealand, signing the Burnham Declaration.[64] This called for the various leaders to bring about a cease-fire and for an international peacekeeping force to be deployed to Bougainville. In October 1997, the Burnham Declaration was followed up by the signing of the Burnham Truce.[65] This established an immediate truce between the conflicting parties on Bougainville and recommended to all parties that a TMG should be deployed to Bougainville. The technicalities of monitoring the Burnham Truce were resolved in meetings in Cairns, Australia, between the PNG government and the Bougainville factions in November 1997.

At the same time in Cairns, an agreement was signed by the governments of Australia, New Zealand, PMG, Fiji, and Vanuatu regarding the terms under which the TMG would operate. Under the terms of the agreement, the TMG had responsibility for:

- Monitoring the compliance of the parties with the terms of the Burnham Truce
- Promoting and instilling confidence in the peace process
- Providing people in Bougainville with information on the truce agreement and the peace process[66]

It is understood that the latter two responsibilities clearly indicated the need for the conduct of IO to support the TMG's activities. The TMG deployed to Bougainville in December 1997 under New Zealand command. The bulk of the TMG personnel, both military and civilian, were provided by Australia and New Zealand, with some participation from Fiji and Vanuatu. The operation of the TMG and later the PMG will be dealt with in the following paragraphs.

In January 1998, the various parties in the Bougainville conflict signed the Lincoln Agreement. This extended the truce period to April 30, 1998, whereupon a permanent cease-fire would come into effect. At the same time, the TMG would be replaced by the PMG, the terms under which the PMG would operate being contained in the so-called Arawa Agreement, which was an annex to the Lincoln Agreement. Finally, the Lincoln Agreement provided for free and democratic elections for a Bougainville Reconciliation Government (BRG).[67] As the Arawa Agreement came into force on April 30, the TMG was re-roled as the PMG and at the same time, command of the force shifted from New Zealand to Australia, where it has remained since.

In the years since 1998, the peace process in Bougainville has continued apace. In August 2001, the Bougainville Peace Agreement was signed at Arawa, which brought together agreements on a referendum regarding the future of Bougainville, a weapons disposal plan, and arrangements for autonomy for Bougainville. In the course of 2002, the PNG parliament enacted legislation regarding autonomy for Bougainville, though more remains to be done before Bougainville might be truly regarded as achieving autonomy with respect to PNG. Reflecting on the success of the peace process, the strength of the PMG has been reduced from a peak of about three hundred to a present strength of about fifty. The PMG remains comprehensively engaged in supporting the peace process, with its current activities focusing on weapon disposal.

IO Contributions to Operation BELISI

Having provided the background to Operation BELISI, attention will now be focused on how IO has contributed to the operation. However before doing this, it is necessary to provide a brief overview of the structure and operations of the PMG as shown in Diagram 6-5. The discussion which follows is based on the structure of the PMG at its peak strength from 1998 to 2001. The PMG is divided into a headquarters (with supporting elements) located at Arawa, and a number of monitoring teams (MT) or liaison teams (LT), each of which are allocated a distinct area of operations (AO). Within their respective AO, each MT/LT conducts regular patrols to monitor the peace agreement and to maintain contact with the local population.

As was noted, IO plays a vital role in the achievement of the PMG mission, as mandated by the Arawa Agreement. IO achieves this by the provision of information about the peace process to the various parties on Bougainville.[68] Furthermore, as an unarmed force, the PMG is critically dependent upon the support and goodwill of the people of Bougainville. The principle means by which the PMG achieves this popular support and performs this role is by the dissemination of information products produced by the military information support team (MIST).[69] The primary responsibility of this team is the production of a variety of printed media,

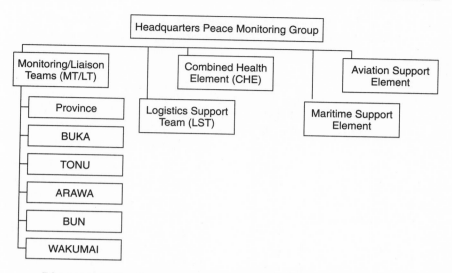

Diagram 6-5 Structure of the Peace Monitoring Group.

of which the most notable is the newspaper titled *Nius Blong Peace* (Peace News), which is supplemented by the glossy monthly magazine *Rot I Go Long Peace* (The Road to Peace).[70] In addition to these two printed products, the MIST is also responsible for the development of a wide variety of other products, including other printed products such as posters and handbills and products such as T-shirts, hats, and soccer balls bearing messages.[71]

Beyond printed products, the MIST has also developed radio scripts and a cassette featuring local music. Collectively, these products support the achievement of the PMG's objectives, as mandated by the Arawa Agreement. They have also been used for information campaigns centered on themes to support the peace objectives, including preventing domestic violence against women as well as the production and consumption of *hombru* (home-brew liquor).[72] Though the MIST does perform some dissemination of the products that it produces, the bulk of these efforts fall to the MTs, from both their static locations, as well as during their patrols. In addition to disseminating MIST product, the MTs were also able to provide feedback to the MIST on the effectiveness of their products. The products produced by the MIST have been well received by all the factions on Bougainville and represent a key means by which the PMG has performed its mission. More than that, it could be argued that Operation BELISI represents an example of IO being the main effort of an operation, with other military elements being in support of these operations. IO has been the means by which the Bougainville population has been kept informed of developments in the peace process and its support for the PMG maintained.

East Timor—Background

The second case study on the operational employment of Australian IO doctrine is the Australian led peace-enforcement mission to East Timor, code-named Operation STABILISE, as shown in Diagram 6-6. The roots of the present U.N. intervention in East Timor date back to 1974.[73] Following the collapse of the government in Portugal, civil war erupted in East Timor between factions favoring independence and those supporting integration with Indonesia. In December 1975, following the withdrawal of the Portuguese administration, Indonesia intervened militarily in the territory, and the following year they integrated East Timor as their twenty-seventh province, an act which was not recognized by the United Nations. Intermittent guerrilla conflict continued off and on between the Indonesian security forces and pro-independence groups (principally FRETLIN and its armed wing, FALINTIL) throughout the intervening period until the late 1990s. Concurrent with the ongoing guerrilla conflict were extensive human rights abuses conducted by the Indonesian security forces against the East Timorese population. Estimates of the total number of deaths in East Timor for a twenty-three-year period ending in 1999 range well above the 100,000 mark, out of a pre-1976 population of about 680,000 for the territory.

The 1997–1998 economic collapse in Asia sparked political turmoil throughout Indonesia. In mid-1998 this culminated in the forced resignation

Diagram 6-6   East Timor's Location Relative to Indonesia.

of President Suharto with replacement B. J. Habibie, prior to the outcome of general elections to be held later in 1999. As part of the measures to deal with political disorder, Habibie floated the idea of an autonomy ballot for East Timor, see diagram 6-6. This eventually led to the signing of the May 5, 1999 agreement between Indonesia and Portugal for the conduct of a ballot on the future status of East Timor. Despite ongoing sporadic violence and intimidation by pro-Indonesian militias, some 446,000 voters registered to take part in the ballot. This culminated in the actual ballot on August 30, 1999, with results announced on September 3, 1999. Some 78.5 percent of the East Timorese electorate voted against special autonomy within Indonesia, or in other words, for East Timorese independence.

In the immediate wake of this announcement, widespread violence and destruction was instigated by pro-Indonesian militia groups. Thousands of East Timorese were killed, and well over 150,000 were displaced from their homes. Despite the declaration of a state of emergency, Indonesian security forces proved either unwilling or unable to stem the bloodshed. As international pressure mounted, evacuation operations were mounted to rescue United Nations personnel and some East Timorese from the unfolding carnage. On September 15, 1999, the U.N. Security Council adopted Resolution 1264, which authorized the deployment of the Australian-led INTERFET force to East Timor in order to restore peace and security.

INTERFET deployed into East Timor on September 20 as Operation STABILISE, and immediately commenced securing the immediate vicinity of Dili, as shown in Diagram 6-7.[74] By late October, the INTERFET force had taken control of all of East Timor, including the Oecussi enclave located in the middle of West Timor. During the subsequent months, INTERFET restored peace and security to East Timor, though the prospect of armed clashes with pro-Indonesian militia infiltrating from West Timor remained a constant threat. In mid-February 2000, INTERFET commenced the hand-over of its responsibilities to a United Nations force and completed the hand-over on February 28, 2000. Operation STABILISE was at an end.

IO Contributions to Operation STABILISE

Having provided the background to Operation STABILISE, we now focus on the conduct of IO during the operation. However, it must be noted that as of August 2002, much of the details pertaining to the planning and conduct of Operation STABILISE still remain outside the public domain. Not the least hidden, in this respect, are the aspects relating to the planning and conduct of IO during the operation.[75] Accordingly, this account of the

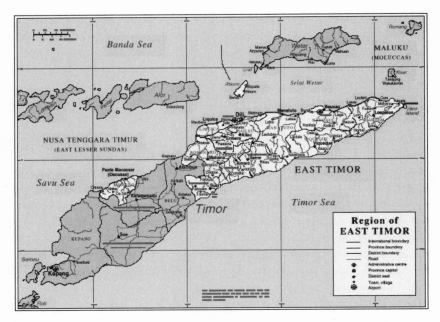

Diagram 6-7   The Region of East Timor.

role played by IO in Operation STABILISE is far from complete, and will concentrate on that one element of IO that, of necessity, is in the public eye—namely public information (PI).

From the information that is readily available from open sources, it is clear that IO played a significant role in the overall conduct of Operation STABILISE. At a public address delivered in April 2000, (then) Major General Peter Cosgrove, the Commander of INTERFET noted:

> ". . . the military operation plainly had an IO quotient to it. By that I mean that our military operations to provide a peaceful and secure environment in which the U.N. could conduct humanitarian assistance and nation building activity were to be seen in two dimensions: what we were actually doing and achieving on the ground; and what we were perceived as doing, its relevance, proficiency and legitimacy."[76]

In his address, Cosgrove divided the parties whose perceptions were critical to the success of Operation STABILISE into four broad groups:

- The individual nations comprising the INTERFET coalition
- The INTERFET coalition itself and the broader international community

- Nations whose view of the INTERFET coalition mission, composition, and leadership Cosgrove termed as "jaundiced"
- Parties within East Timor

These stakeholders and their perceptions were the crucial elements that the INTERFET force needed to address, including the last group which was comprised of the East Timorese population, the various U.N. agencies, and NGOs.[77]

From the outset, Cosgrove assessed that there would be an IO campaign by interests opposed to the INTERFET mission, such as the pro-integrationist militia groups, to discredit the coalition operation. From this assumption it followed that such an IO campaign would have to be countered vigorously and effectively as a highest priority. Therefore, in common with the leaders of many past coalition operations, Cosgrove identified that the center of gravity of the INTERFET coalition was the maintenance and legitimacy of the coalition itself.[78] Furthermore, he also identified that the chief means by which the coalition center of gravity might be targeted was via adversary IO, namely the misinformation and propaganda disseminated via the global electronic media.[79] Noting the media attention that the East Timor operation had generated, Cosgrove chose to embrace and encourage the presence of the global media in East Timor, rather than merely accept or acquiesce to its presence.[80] He emphasized that the INTERFET force would be transparent, accountable, available, and very pro-active in dealing with the global media. Cosgrove felt that the most effective way of countering adversary propaganda and disinformation was to invite open scrutiny of INTERFET personnel, activities, and operations by the media and to allow audiences to assess the lies for what they were.

A key example of this policy in action was the INTERFET response to claims that militia groups had infiltrated in strength deep into East Timor from West Timor. In the second week of October 1999, television footage was broadcast around the globe purporting to show militia leader Eurico Guterres, with 150 militia personnel, in the vicinity of Liquica.[81] Guterres claimed to have crossed from West Timor into East Timor, and then to have driven some 100 kilometers to Liquica, without sighting or being sighted by any coalition INTERFET troops. The INTERFET commander dealt directly with the media and countered the claims by highlighting discrepancies and improbabilities in the militia leader's account, thereby extinguishing the credibility of Guterres' claims. Had INTERFET not engaged the global media as thoroughly as it did, it is unlikely that the countering of such claims would have been nearly as effective. In short, INTERFET's truthfulness and openness in dealing with the media was its greatest asset in countering militia propaganda.

Although this account of IO during Operation STABILISE has focused on PI activities, this is primarily the result of the lack of open-source information on other IO activities that have been undertaken in East Timor. That said, the significance of PI in establishing and maintaining the credibility of the coalition in the face of militia propaganda and disinformation should not be undervalued. For if it is the case that the legitimacy of the INTERFET coalition was its center of gravity, then it follows that efforts taken to protect that center of gravity are a core rather than peripheral part of warfighting. In this regard, the final words here will be left to Cosgrove who said, "I cannot stress enough this aspect of IO in its crucial contribution to a successful coalition mission. In this area of nurturing your constituencies, you can be figuratively just as damaged by a headline as a bullet."[82]

Lessons Learned and Directions Forward
Having now reviewed the employment of IO by Australia in two recent contingencies, an assessment will now be made of the effectiveness of the Australian approach to IO and the means by which this can be improved. In general terms, it can be clearly stated that Australia's operational employment of IO has been effective. In both Operation BELISI and STABILISE, IO contributed significantly, if not critically, to the success of both operations. In this sense, then, both of these operations can be seen as validating the Australian approach to IO. That said, certain caveats apply. In both cases, Australian forces were working in fairly unsophisticated, low intensity operational environments. In neither instance was the ADF faced with an adversary of the sophistication of either Serbia or Iraq.[83] The full range of IO-related capabilities was never fully employed. So leaving this issue to the side for the moment, how can Australia's ability to conduct IO be further enhanced?

With respect to acquiring new capabilities, significant effort is already underway. As was outlined earlier, the unclassified Defence Capability Plan includes a number of initiatives to acquire new EW, CND/IA, and other information system security capabilities. These will all markedly enhance the ADF's ability to conduct IO in a technically sophisticated environment. Having acquired such new or enhanced capabilities, the challenge then remains as to how to incorporate them into the ADF's force structure, and how to employ them operationally.

The guidance required to meet this challenge is provided by doctrine. The current draft doctrine will be published as an ADDP in late 2002, and the final product will incorporate the lessons learned from Operations BELISI and STABILISE, as well as from observations of other nations' efforts. A particular issue for consideration in this regard is the conduct of IO within a coalition framework. It has been claimed by some commentators that most future military operations will be conducted by impromptu coalitions rather than traditional allies.[84] Present Australian and U.S. IO doctrine barely touch on the issues involved in conducting IO in such a setting. This is a matter of great importance, particularly in the context of the ongoing war on terrorism.

A final issue to note is the role of IO in the conduct of national security affairs as a whole. The focus of this chapter has been on the employment of IO by the ADF at the tactical and operational levels of war. But the ADF (or any other military) can only conduct effective military IO within a whole-of-government (and indeed, whole-of-nation) framework. As just one example, the ADF (like any other modern military) is critically dependent upon the overwhelmingly civilian and commercial NII. In turn, this matter is closely related to the broader issue of an integrated and holistic approach to national security. At present, these matters remain in their relative infancy in Australia. But in common with many other countries, these issues are presently being considered in the context of reforming national security arrangements to meet the security challenges of the post-September 11th world.

Summary

As shown in this chapter, there are a number of nations other than the United States that have conducted IO missions or operations and are currently writing doctrine, but this text has concentrated on the few preceding examples. First, in the case of Russia and China, these two nations are the closest peer competitors to the United States and any new developments by these countries are watched with great interest by Western military forces. Second, the operations in Kosovo gave the United States and the coalition a chance to use the new IO doctrine against a savvy adversary, but unfortunately the advantages inherent in this warfare area were not properly utilized. An opportunity missed, Operation Noble Anvil also demonstrated how a supposedly less sophisticated force (i.e., Serbia) was able to successfully deceive and manipulate coalition forces during this seventy-eight-day campaign. Finally, the section on Australia, including its doctrinal and operational use of IO over the last few years, is very illuminating for a number of reasons. A small force, the ADF is a great example of how the asymmetric capabilities inherent in IO are used by forces other than the United States. Maybe it is these nations that will have the greatest impact on IO in the future. It is our goal that this section demonstrates, in an unclassified format, some of the unique characteristics that make up the power embedded in IO.

Conclusion

What Is the Future of Information Operations?

"We didn't give up when the Germans bombed Pearl Harbor . . ."

Bluto

Information is power, and how a nation uses that power determines how effective a country may be in influencing the world politic. Unlike in the past when the elements of power included only military, economic, and diplomatic factors, in the twenty-first century, information is rapidly assuming a place of primacy in the conduction of foreign policy. It can be a force multiplier, a decision-tool, a central theme for an offensive campaign, and so much more. But to be useful, information must be understood for what it truly is—a weapon, and if not used correctly, it can backfire just like any other kinetic device in the arsenal.

The use of information to affect public opinion has a long and varied history within world politics. Often it was the government or leadership elite that could control that information, thereby exercising power over their people. Yet the tremendous advances in technology in the computer and telecommunications fields over the last decade have shattered their monopoly of control over information. In addition, the merging of these formerly separate areas has given the power to use information to a much greater audience. This in turn has forced the government to work harder to control the dissemination and ultimately the use information as an element of power.

Yet in reality, the government can no longer control information. This is because it does not own the sources or the means of delivery of information to our modern society. Which of course begs the question, if you cannot control information, can you really control power? There are many other organizations outside the government that now have a much greater influence on the flow of information, and it is now the government more often than not that is on the defensive. Because it cannot control the information, it must therefore react, and because the government is a bureaucracy, it cannot act fast enough to stay on the offensive.

Therefore, information operations is changing the way in which the world's militaries and governments conduct business. This includes military deterrence and peacekeeping operations, foreign policy, and also worldwide economic development. No longer can these missions be conducted in isolation, and so it is imperative that the organizational structure adapt to these changing circumstances. That is precisely why so much turmoil exists in the governmental architecture today. Will these changes even-out as IO matures as a warfare area? Perhaps, but it remains to be seen as the United States and other nations continue to develop the weapons and capabilities of information operations.

The use of information as a weapon and force multiplier is not likely to go away with a change in an administration or government. A fundamental shift has occurred and the world is now in the midst of a revolution, a new era in which information is now the most fungible of powers, and whoever uses it to their best advantage will emerge victorious. Unfortunately, the fact that the dynamics of power have greatly changed is not widely recognized at this time. As we questioned at the very beginning of this book, would you recognize a revolution if you were in it? Information operations has forever changed the method of conducting warfare. It is hoped that some of the more crucial concepts of this new warfare area have been covered adequately by this book and will be useful to you in your various operations.

The authors hope that you enjoyed this publication. This book was meant to serve not only in a teaching role, but also as an update for the millennium. We chose to highlight the critical time period from 1989 to 2004, with the significant events of *ER '97* to the incorporation of IO into military doctrine with JV 2020 as the highlights. In these last several years, much has changed within the IO community, both in organizational activity as well as doctrinal publications. However the tragic events of September 11, 2001 have forever changed the concept of IO, and have shown the true power of information. We have tried to address these issues as well as looking at what else is happening around the world concerning IO, concentrating on Russia, China, and Australia because that is where a great deal of activity has occurred.

To conclude, the editors and all the contributors hope you enjoyed this book. We are using this book as a primary reference source for our students, and, therefore, if you see mistakes or upgrades that need to be made, please feel free to contact us at *jciws-iw@jfsc.ndu.edu.* Our plan is to continue to use this publication in the future, so any help you can give us would be greatly appreciated. Once again, thanks, and we hope you enjoyed reading this book.

Endnotes

Introduction

[1] Associated Press, "VOA Airs Interview with Taliban Leader," (26 September 2001): *Washington Post*, p. 1.

[2] PDD-68 was addressed to the Defense, State, Justice, Commerce, and Treasury departments and the CIA and FBI.

[3] Floodnet is a JAVA applet that causes constant search queries of a site by the search engine every nine seconds. It monopolizes computer processor unit time and resources that may cause the server to overload.

[4] *Hactivism* is the use of hacker or cyber-attacks to promote activism in a particular cause.

[5] RAND, *The Zapatista Social Netwar in Mexico* (Washington, D.C.: RAND Corporation, 1996), 45.

Chapter 1

[1] Phil Alden Robinson (director), *Sneakers*, 1992.

[2] John Arquilla and David Ronfeldt, *The Emergence of Noopolitik: Toward an American Information Strategy* (Santa Monica, Ca.: RAND, 1999).

[3] John Arquilla and David Ronfeldt, "Looking Ahead: Preparing for Information-Age Conflict," in *In Athena's Camp: Preparing for Conflict in the Information Age,* eds. John Arquilla and David Ronfeldt (Santa Monica, Ca.: RAND, 1997), 441.

[4] Hans J. Morganthau, *Politics among Nations: The Struggle for Power and Peace* (New York: Alfred A. Knopf, 1967), xviii.

[5] Joseph S. Nye and William A. Owens, "America's Information Edge," *Foreign Affairs* 75 (March/April 1996): 22.

[6] Walter B. Wriston, "Bits, Bytes and Diplomacy," *Foreign Affairs* 76 (September/October 1997): 175.

[7] Barbara Haskell, "Access to Society: A Neglected Dimension of Power," *International Organization* 34 (Winter 1980): 94; Joseph S. Nye, *Bound to Lead: The Changing Nature of American Power* (New York: Basic Books, Inc., 1990), 8; Eliot A. Cohen, "A Revolution in Warfare," *Foreign Affairs* 75 (March/April 1996): 52; Robert O. Keohane and Joseph S. Nye, "Power and Interdependence in the Information Age," *Foreign Affairs* 77 (September/October 1998): 81; Joseph S. Nye and William A. Owens, "America's Information Edge,"

Foreign Affairs 75 (March/April 1996): 20; Walter B. Wriston, "Bits, Bytes and Diplomacy," *Foreign Affairs* 76 (September/October 1997): 172; Richard N. Rosecrance, *The Rise of the Virtual State: Wealth and Power in the Coming Century* (New York: Basic Books, Inc., 1999), 16; John Arquilla and David Ronfeldt, "A New Epoch—And Spectrum—Of Conflict," in *In Athena's Camp: Preparing for Conflict in the Information Age,* eds. John Arquilla and David Ronfeldt (Santa Monica, Ca.: RAND, 1997), 7.

[8]David C. Gompert, *Right Makes Might: Freedom and Power in the Information Age* (Washington, D.C.: National Defense University, 1998), 5.

[9]Joint Publication 3-13, *Information Operations* (Washington, D.C.: Government Printing Office, 9 October 1998).

[10]Joint Chiefs of Staff, *Joint Vision 2010* (Washington D.C.: Government Printing Office, July 1996), 69.

[11]The first known use of information warfare was in a briefing title and concept written by Dr. Tom Rona (then of Boeing) for Andrew Marshall, May/June 1976.

[12]Neil Munro, *The Quick and the Dead: Electronic Combat and Modern Warfare* (New York: St. Martin's Press, 1991), 173.

[13]U.S. Department of Defense, DEPSECDEF MEMORANDUM, *Strategic Concept for Information Operations (IO),* by John J. Hamre (Washington, D.C.: 14 May 1999).

[14]Ibid, art. I, sec. 8 (11).

[15]National Security Council Organization Chart (accessed 14 June 1999), available from http:www.whitehouse.gov/wh/eop/nsc. Members include the chairman of the Joint Chiefs of Staff, director of Central Intelligence, secretary of Treasury, assistant to the president for national security affairs, assistant to the president for economic security, United States representative to the United Nations, and the president's chief of staff, in addition to the attorney general and the head of the Office of National Drug Control Policy.

[16]This includes the new Department of Homeland Security.

[17]These include: Council of Economic Advisors, Council on Environmental Quality, National Economic Council (NEC), National Security Council (NSC), Office of Administration, Office of the First Lady, Office of Management and Budget (OMB), Office of National Drug Control Policy (ONDCP), Office of Science and Technology Policy (OSTP), White House Office for Women's Initiatives and Outreach, President's Foreign Intelligence Advisory Board, and the United States Trade Representative (USTR).

Three of these offices, the NSC, OMB, and OSTP, have a direct interest in IO policy, whereas others, such as the NEC, ONDCP, and USTR, are involved on an ancillary basis.

[18]Specific duties assigned to the NSC by law include:

- "The function of the Council shall be *to advise the President* with respect to the integration of domestic, foreign, and military policies relating to the national security . . ."

- ". . . performing such *other functions the President may direct* for the purpose of more effectively *coordinating the policies and functions* of the departments and agencies of the Government relating to the national security . . ."
- ". . . *assess and appraise* the objectives, commitments, and risks of the United States . . ."
- ". . . *consider policies* on matters of common interest to the departments and agencies of the Government concerned with the national security . . ."

[19]This is spelled out explicitly in the NSC's list of official functions: provide information and policy advice to president; manage the interagency policy coordination process; monitor implementation of presidential policy decisions; crisis management; support negotiations; articulate president's policies; liaison with Congress and foreign governments; coordinate summit meetings and national security-related trips.

[20]President, Executive Order 13073 "President's Council on Year 2000 Conversion" (4 February 1998).

[21]President, Amendment to Executive Order 13073 (14 June 1999).

[22]President, Executive Order 12881 "National Science and Technology Council (NSTP)" (23 November 1993); and Executive Order 12882 "President's Committee of Advisors on Science and Technology" (23 November 1993).

[23]Ibid.

[24]President, Presidential Decision Directive 29 "Security Policy Coordination . . ." (16 September 1994).

[25]U.S. Department of Defense, Under Secretary of Defense for Policy and Assistant Secretary of Defense for Command, Control, Communications and Intelligence, Joint Memorandum, *Direction to the Staff: Seamless Integration between USD(P) and ASD(C3I) regarding Information Operations,* by Walter B. Slocombe and Arthur L. Money (Washington, D.C.: 1999).

[26]President, Executive Order 12333 "United States Intelligence Activities . . ." (4 December 1981).

[27]President, National Security Telecommunications and Information Systems Security Committee Document (NSTISSCD) 503 "Incident Response and Vulnerability Reporting for National Security Systems" (30 August 1993).

[28]President, National Security Decision 42 "National Policy for the Security of National Security Telecommunications and Information Systems" (5 July 1990).

[29]U.S. Department of Defense, The Joint Staff, "Information Assurance: Legal, Regulatory, Policy and Organizational Considerations" (August 1999), A-88.

[30]Ibid, 5-2.

[31]President, Interdepartmental Committee on Communications (26 October 1921), updated (21 August 1963) and by Executive Order 12472 "Assignment of National Security and Emergency Preparedness Telecommunications Functions" (3 April 1984).

[32]President, Executive Order 12382 "National Security Telecommunications Advisory Committee" (13 September 1982), continued by Executive Order 12610 (30 September 1987).

[33]Department of Defense, "Information Assurance," A-79.

[34]To get more information on IATAC and which products are currently available, you can contact its director, Robert Lamb, at (703) 289-5454 or iatac@dtic.mil.

[35]U.S. Department of Defense, Joint Publication 3-13, "Joint Doctrine for Information Operations" (9 October 1998).

[36]U.S. Department of Defense, Chairman, Joint Chiefs of Staff Instruction (CJCSI) S3210.01A "Joint Information Operations Policy (U)" (5 November 1998).

[37]U.S. Department of Defense, Joint Chiefs of Staff, "Unified Command Plan Changes 1999" (accessed on 2 March 2000), available at http://www.defenselink.mil/specials/unified/planchanges1.html.

[38]Ibid.

[39]U.S. Department of State, *Reform and Restructuring Act* (30 December 1998).

[40]President, Presidential Decision Directive 68 "International Public Information" 68 (30 April 1999).

[41]President, Presidential Decision Directive 63 "Critical Infrastructure Protection" (22 May 1998).

[42]President, President's Commission on Critical Infrastructure Protection (PCCIP), *Critical Foundation: Protecting America's Infrastructures* (13 October 1997).

[43]Ibid.

[44]Originally formed as the National Bureau of Standards in 1901, NIST was renamed and reorganized by the Omnibus Trade and Competitiveness Act of 1988, U.S. Code P.L. 100-418, 102 Stat. 1107.

[45]U.S. Department of Commerce, National Institute of Standards and Technology Authorization Act (1989).

[46]Computer Security Act of 1987 (P.L. 100-235) and Information Technology Management Reform Act of 1996, National Defense Authorization Act of Fiscal Year 1996 (10 February 1996) P.L. 104-106.

[47]U.S. Department of Defense, National Security Agency, and Department of Commerce, Letter of Partnership "National Security Agency and National Institute of Standards and Technology" (22 August 1997).

[48]NTIA was created by Reorganization Plan Number 1 (1977) and implemented by Executive Order 12406 "Relating to the Transfer of Telecommunications Functions" (25 March 1978).

[49]PDD-63.

[50]Ibid.

[51]An NGO is a transnational organization of private citizens that maintains a consultative status with the Economic and Social Council of the United Nations. NGOs may be professional associations, foundations, multi-national businesses, or simply groups with a common interest in humanitarian assistance activities (development or relief).

Chapter 2

[1]Sun Tzu, *The Art of War,* 400–320 B.C., quoted in JP 2.0, p. IV–14.

[2]JP 3-13, pp. vii–viii.

[3]JP 3-13, GL-5.

[4]Graphically, the fundamental aspects of intelligence (relevant information) and command, control, communications, and computers (information systems) in supporting information operations are shown in Figure 1-2, "Information Operations Across Time," in JP 3-13, p. I-4.

[5]JP 3-13, p. I–5.

[6]The problem with a smorgasbord approach to providing intelligence products is that it increases the likelihood of security breaches and affects a command's operations security (OpSec) posture when "need to know" is thrown by the wayside.

[7]Carl Von Clausewitz, *On War,* 1832, quoted in JP 2.0 *Joint Doctrine for Intelligence Support to Operations* (5 May 1995): IV-1.

[8]General Colin Powell, quoted in JP 2.0, p. IV–7.

[9]In most cases, intelligence officers will generate these on behalf of a commander and then seek approval after the fact if they have not received specific guidance.

[10]JP 2.0, p. III–4.

[11]DRAFT Revision of JP 2.0, Second Final Coordination (13 July 1998): II-1.

[12]JP 2.0 *Doctrine for Intelligence Support* (9 March 2000): II-3.

[13]Different commands have other names (i.e. EuCom calls it the Joint Analysis Center [JAC] and JFCOM calls it the Joint Forces Intelligence Center [JFIC]).

[14]For a complete description of each category, see DRAFT JP 2.0, Second Final Coordination Copy (13 July 1998): II-12 through II-13.

[15]Robert V. Ackerman, "Military Intelligence Looks Within," *SIGNAL* (October 2000): 16.

[16]Compared to the one contained in the earlier version of JP 2.0, p. II–3.

[17]Most intelligence analysts and collectors understand the system and know how to work it to their advantage. For example, one analyst working in scientific and technical intelligence always gave every HumInt report that cited her Critical Informant Rating (CIR) a level "of major significance" simply because the priority of collection against her requirements was so low. Collectors, in turn, who were rated based on the number of evaluations they received "of major significance" learned then to cite her CIR often, knowing they would get top marks simply by asking their sources a few questions related to her intelligence-production tasking.

[18]A case in point is the *Mayaguez* incident which occurred in May 1975. When the Cambodians seized the U.S.-flagged merchant vessel *Mayaguez,* a presidential decision was made to effect a hostage rescue operation. At the time the operation was conducted, there were contradictory intelligence reports on the size of the opposition to be encountered, as well as the location of the captive crew. Furthermore, within the intelligence community, there were no established

procedures to deconflict intelligence discrepancies. The Marines who conducted the assault on Kao Tang Island suffered large numbers of casualties due to the underestimated size of the opposition, while the crew of the *Mayaguez* had been released hours earlier from a different location. See Patrick W. Urey, *The* Mayaguez *Operation,* Center for Naval Analysis Document No. CNS 1085, May 1991, and Lieutenant Colonel (LTC) Theodore H. Mueller, "Chaos Theory: The *Mayaguez* Crisis," U.S. Army War College, Carlisle, Pa., March 1990.

[19] *A Consumer's Guide to Intelligence,* Central Intelligence Agency, Document No. PAS 95-00010 (July 1995): 39.

[20] *CJCS Report on the Roles, Missions, and Functions of the Armed Forces of the United States* (February 1993) quoted in DRAFT JP 2.0, p. IV–1.

[21] DRAFT JP 2.0, p. IV–8.

[22] JP 2-01.3 *Joint Tactics, Technique and Procedures for Joint Intelligence Preparation of the Battlespace* (24 May 2000).

[23] Ibid.

[24] Items known as OCOKA: observation, cover and concealment, obstacles, key terrain, and avenues of approach.

[25] In 1995, when NATO bombed Serbia to pressure Milosevic to back off supporting Bosnian Serbs, he did capitulate after a couple of days of intense air strikes. During the Kosovo conflict, intelligence analysts failed to calculate the psychological, historical, and cultural ties of Serbs to Kosovo that were much stronger than those to Bosnia-Herzegovina. In addition, Milosevic was willing to risk his country's survival (and his political survival) over Kosovo.

[26] JP 3-13, pp. II–11 through II–13.

Chapter 3

[1] JP 3-13, p. I–10.

[2] Ibid, p. III–1.

[3] Former Secretary of Defense Dick Cheney, quoted in *Concept for Future Joint Operations: Expanding Joint Vision 2010,* The Joint Staff (May 1997): 14.

[4] JP 3-13, p. GL–5.

[5] A precedent-setting case occurred on September 9, 1998, when the Pentagon, having advanced notice of a cyber-attack, actively shut down the Internet browsers of those logging in to participate in the attack. See George Seffers, "Thwarted Hackers Call Pentagon Actions 'Offensive,'" *Army Times* (October 12, 1999).

[6] Joint Publication 3-13, p. I–9.

[7] *The Insider Threat to U.S. Government Information Systems,* NSTISSAM INFOSEC/1-99 (July 1999), National Security Telecommunications and Information Systems Security Committee.

[8] Director Richard Clarke, NSC, CBS "60 Minutes" (9 April 2000).

[9] CJCS MEMO CM-510-00 (10 March 1999).

[10] *Information Assurance Defense In Depth,* JCS (2000).

[11]Deputy Secretary of Defense, *Implementation of the Recommendations of the Information Assurance and Information Technology Integrated Process Team on Training, Certification and Personnel Management in the Department of Defense* (14 July 2000).

[12]Office of the Assistant Secretary of Defense Command, Control, Communications, and Intelligence, DoD CIO Annual Information Assurance Report (April 2000).

[13]For more information on these different organizations and their efforts, see the following website: http://www.biometrics.org/.

[14]General Henry H. Shelton, Chairman, Joint Chiefs of Staff, *The Transatlantic Commitment* at "NATO at 50" conference, London (8 March 1999).

[15]Bruce Hoffman, *Inside Terrorism* (New York: Columbia University Press, 1998) 13.

[16]Walter Laqueur, *Terrorism* (Boston: Little, Brown and Co., 1977), 7.

[17]U.S. Department of Justice, *Terrorism in the United States 1997* (Washington, D.C.: Federal Bureau of Investigation, 1998), ii. In their book *Political Terrorism,* Schmid and Jongman, 1983, New Brunswick: Transaction Books, cited 109 different definitions of terrorism, which they obtained in a survey from leading terrorism scholars. From these definitions, the authors isolated the following recurring elements, in order of their statistical appearance in the definitions: violence force 83.5 percent of the definitions; political 65 percent; fear, emphasis on terror 51 percent; threats 47 percent; psychological effects and anticipated reactions 41.5 percent; discrepancy between the targets and the victims 37.5 percent; intentional, planned, systematic, organized action 32 percent; methods of combat, strategy, tactics 30.5 percent.

[18]Cited in Richard W. Leeman, *The Rhetoric of Terrorism and Counterterrorism* (New York: Greenwood Press, 1991), 11.

[19]Walter Laqueur, *The Age of Terrorism* (Boston: Little, Brown, 1987), 306.

[20]Ibid.

[21]Cited in Richard W. Leeman, *The Rhetoric of Terrorism and Counterterrorism* (New York: Greenwood Press, 1991), 19.

[22]Secretary of State Madeleine K. Albright, American Legion Convention (9 September 1998).

[23]For a more in-depth discussion of the classification of these seven types of terrorist groups please see *Counter-Terrorism Information Operations Case Study* by The Rendon Group, August 1999. Independently Published Consultant Firm.

[24]Carnes Lord and Frank Barnett, eds., *Political Warfare and Psychological Operations: Rethinking the U.S. Approach* (Washington, D.C.: National Defense University Press, 1988), 160.

[25]U.S. President. *Presidential Decision Directive 39—US Counter-Terrorism Policy,* Executive Office of the President, Washington, D.C., 21 June (initially classified Secret, declassified to Unclassified with deletions on 24 January 1999). HTML document available from: http://www.fas.org/irp/offdocs/pdd39.htm. Accessed 27 August 2003.

[26]Cordesman, A. H. & Cordesman, J. G. (2002)—*Cyber-threats, Information Warfare and Critical Infrastructure Protection: Defending the US Homeland,* Praeger, Westport, Connecticut, pp. 1–2.

[27]PDD-39; Vatis, M. (1998)—*Statement for the record on Infrastructure Protection and the Role of the National Infrastructure Protection Center,* before the Senate Judiciary Sub-Committee on Technology, Terrorism and Government Information, Washington, D.C., 10 June. HTML document available from: http://www.fbi.gov/congress/congress98/vatis0610.htm. Accessed 27 August 2003.

[28]U.S. President. *Executive Order 13010—Critical Infrastructure Protection,* Executive Office of the President, Washington, D.C., 17 July.

[29]National Communication System. *The Electronic Intrusion Threat to National Security and Emergency Preparedness Telecommunications: An Awareness Document,* National Communications System, Washington, D.C., August 1993; Information Infrastructure Task Force. *NII Security: The Federal Role—* Draft Report, Information Infrastructure Task Force, Department of Commerce, Washington, D.C., 5 June 1995; Science Applications International Corporation. *Information Warfare: Legal, Regulatory, Policy and Organizational Considerations for Assurance,* Report to the Joint Staff, Science Applications International Corporation, McLean, Virginia, 4 July 1995; *Report of the Defense Science Board Task Force on Information Warfare—Defense,* Office of the Under Secretary of Defense for Acquisition and Technology, Washington, D.C., November 1996; Patrick Moynihan. *Report of the Commission on Protecting and Reducing Government Secrecy,* Senate Document 105-2, U.S. Government Printing Office, Washington, D.C., 1997; Gregory Rattray. *Strategic Warfare in Cyberspace,* MIT Press, Cambridge, Massachusetts, 2001.

[30]Vatis, 1998.

[31]*Critical Foundations: Protecting America's Infrastructures,* Report of the President's Commission on Critical Infrastructure Protection, Washington, D.C., October 1997.

[32]J. D. Moteff, *Critical Infrastructures: Background, Policy and Implementation,* Congressional Research Service, The Library of Congress, Report No. RL30153, Washington, D.C., 7 August 2003, p. 4.

[33]Vatis, 1998.

[34]U.S. President. *Summary of PDD-62 and PDD-63,* Executive Office of the President, Washington, D.C., 22 May.

[35]*Fact Sheet—Combating Terrorism: Presidential Decision Directive 62,* Executive Office of the President, Washington, D.C., 22 May 1998.

[36]The Clinton administration announced Presidential Decision Directive 39 on June 21, 1995, which stood as the U.S. counterterrorism policy and left a great deal of implemental discretion to federal agencies. It is a very simplified, broad approach to combating terrorism.

[37]WMD means chemical, biological, or nuclear weapons employed by terrorists.

[38]OMB Annual Report 1999.

[39]Office of the Press Secretary, *Fact Sheet: Countering Terrorism,* PDD-62 (Washington: White House, May 22, 1998).

⁴⁰*White Paper—The Clinton Administration's Policy on Critical Infrastructure Protection: Presidential Decision Directive 63,* Executive Office of the President, Washington, D.C., 22 May 1998.

⁴¹*Defending America's Cyberspace—National Plan for Information Systems Protection 1.0—An Invitation to Dialogue,* Executive Office of the President, Washington, D.C., January 2000.

⁴²Moteff, 2003, 5.

⁴³Many of the functions of this new office are very similar to those of the National Domestic Preparedness Office. Both aim to improve coordination efforts among federal counterterrorism agencies as well as to review existing federal programs. However, this new office would not have the authority to affect the daily operations of such agencies as the FBI, DoJ, or the Department of State. This is the same problem that the National Coordinator for Security, Infrastructure Protection, and Counterterrorism within the National Security Council faces: the lack of authority to affect actual counterterrorism operations and expenditures.

⁴⁴U.S. President. *Executive Order 13228—Establishing the Office of Homeland Security and the Homeland Security Council,* Executive Office of the President, Washington, D.C., 8 October 2001.

⁴⁵Ibid, 8.

⁴⁶U.S. President. *Executive Order 13231—Critical Infrastructure Protection in the Information Age,* Executive Office of the President, Washington, D.C., 16 October 2001.

⁴⁷Moteff, 2003, 10.

⁴⁸*National Strategy for Homeland Security,* Office of Homeland Security, Washington, D.C., 16 July 2002.

⁴⁹*National Strategy to Secure Cyberspace—For Comment Draft,* President's Critical Infrastructure Protection Board, Washington, D.C., 18 September.

⁵⁰B. Berkowitz & R. W. Hahn. 'Cybersecurity—Who's watching the store?' *Issues in Science and Technology,* Vol. 19 No. 3 (Spring 2003), pp. 55–62.

⁵¹Moteff, 2003, 11.

⁵²DHS Organization, Department of Homeland Security, Washington, D.C. HTML document available from: http://www.dhs.gov/dhspublic/display?theme=9&content=1075. Downloaded 27 August 2003.

⁵³*Road Map for National Security: Imperative for Change,* The Phase III Report of the U.S. Commission on National Security/21st Century, U.S. Commission on National Security/21st century, Washington, D.C., 15 February 2001, p. 11–20; Moteff, 2003, 8–9.

⁵⁴*National Strategy to Secure Cyberspace,* Executive Office of the President, Washington, D.C., February 2003; *National Strategy for the Physical Protection of Critical Infrastructures and Key Assets,* Executive Office of the President, Washington, D.C., February 2003; *National Strategy for Combating Terrorism,* Executive Office of the President, Washington, D.C., February 2003.

⁵⁵U.S. President—*Progress Report on the Global War on Terrorism,* Executive Office of the President, Washington, D.C., September, 2003, p. 5.

⁵⁶Ibid, pp. 17–21.

[57]GAO/NSIAD-1999-135, *Combating Terrorism,* p. 10.

[58]James Notter and John McDonald, *Building Regional Security: NGOs and Governments in Partnership* (Washington, D.C.: Institute for Multi-Track Diplomacy, July 1998), 1.

[59]Ibid.

[60]Ibid.

[61]U.S. Department of State, *Annual Report on Global Terrorism 1996–1999.*

[62]The White House, *Remarks by the President to the Opening Session of the 53rd United Nations General Assembly* (New York: Department of State, 1998), 1.

[63]Federation of American Scientists, *U.S.-EU Counterterrorism Summit* (May 18, 1998).

Chapter 4

[1]Lieutenant General S. Bogdanov, former chief of the General Staff Center for Operational and Strategic Studies, October 1991, quoted in JP-3-13, p. II–15.

[2]JP 3-13, p. GL–9.

[3]JP 3-13, p. viii.

[4]Interestingly, within Australian IO doctrine this is referred to as "decision superiority."

[5]This observation was made by Brigadier General John Goodman, Chief of Staff, U.S. Southern Command, during an office call in Miami, Florida, September 1998.

[6]Office of the Press Secretary, the White House, *Remarks by the President on Keeping America Secure for the 21st* Century (Washington, D.C.: National Academy of Sciences, January 22, 1999).

[7]U.S. Department of Defense, The Joint Staff, 3-13 (9 October 1998): GL-5.

[8]"Satellite Spying Cited by Johnson," *New York Times* (17 March 1967), accessed from http://webster.hibo.no/asf/Cold_War/report1/williame.html.

[9]White House Press Release, May 1, 2000, "Improving the Civilian Global Positioning System."

[10]U.S. Department of Defense, JP 3-13 *Joint Doctrine for Information Operations* (Washington, D.C.: The Joint Staff, 9 October 1998): GL-7.

[11]Ibid.

[12]The use of the media as a tool of an information campaign has been given a new term in the last decade, namely, *soft power.* First coined in his book *Bound to Lead,* Joseph Nye defined soft power as the capability that you get when someone wants to be like you. He went on in later foreign affairs articles to explain that soft power is the ability to achieve goals through attraction rather than coercion. It works by convincing others to follow or getting them to agree to norms and institutions that produce the desired behavior.

[13]Presidential Decision Directive 56 Managing Complex Contingency Operations (May 1997).

[14]International military information operations are overt/public activities to convey selected information and indicators to foreign audiences to influence their emotions, motives, objective reasoning, and ultimately the behavior of foreign governments, organizations, groups, and individuals.

[15]LTC Chris Pilecki, 6 June.

[16]Interview with Col Jack Summe, 6 June 2000 by the Author, Pentagon.

[17]Franklin Foer, "Flacks Americana—John Rendon's Shallow PR War on Terrorism," *New Republic* (20 May 2002).

[18]Foer, "Flacks Americana."

[19]Jonathen Atter, "Why Hughes Shouldn't Go," *Newsweek* (6 May 2002): 49.

[20]Interview with Tom Timmes, assistant for PSYOP and public diplomacy, ASD-SOLIC (Washington, D.C., April 15, 2002).

[21]Paul Rodriguez, "Goodbye for Now, Tori," *Insight on the News* (10 June 2002), accessed April 14 2003 from http://www.insightmag.com/global_user_elements/printpage.cfm?.

[22]Bob Woodward and Dan Balz, "Aus Trauer Wird Wut," *Der Welt* (16 February 2002), accessed March 7, 2002 from http://www.welt.de/daten/2002/02/16/0216au314625.htx?; Florian Rötzer, "Rumsfield: Pentagon Lügt Nicht," *Die Heise* (21 February 2002), accessed March 15, 2002 from http://www.heise.de/tp/deutsch/special/info/11895/1.html.; Florian Rötzer, "Aus für de Propaganda-Abteilung des Pentagon," *Die Heise* (27 February 2002), accessed March 25, 2002 from http://www.heise.de/tp/deutsch/special/info/11952/1.html.; Martin Van Crevald, "Die USA im Psychokrieg," *Der Welt* (3 March 2002), accessed 1 April 2002 from http://www.welt.de/daten/2002/03/01/0301fo317407.htx?.

[23]U.S. Department of Defense, Defense Science Board on Managed Information Dissemination, October 2001.

[24]Florian Rötzer, "Schon wieder eine Chinesische Mai-Offensive," *Die Heise*, (25 April 2002), accessed 30 May 2002 from http://www.heise.de/tp/deutsch/special/info/12401/1.html.

[25]Interview by the author with LTC Brad Ward (USA), DoD Representative for the IPI Core Group the DoS, 15 April 2002.

[26]Interview by the author with John Rendon, The Rendon Group, Washington, D.C. (16 April 2002).

[27]George W. Downs, ed., *Collective Security beyond the Cold War* (Ann Arbor: The University of Michigan Press, 1997), 23.

[28]U.S. president, White House website. Accessed November 24, 2001 from http://www.whitehouse.gov/response/militaryresponse.html.

[29]Ibid.

[30]John Cloud, "The Manhunt Goes Global," *TIME* (15 October 2001), 52.

[31]Matt Rees, "Jordan on the Terror Trail," *TIME* (26 November 2001), 10.

[32]Ibid.

[33]Edward T. Pound, "The Root of All Evil," *US News and World Report* (3 December 2001), 29.

[34]Op. cit.

[35]General Tommy Franks, "Statement to the Senate Armed Services Committee," U.S. Senate update (7 February 2002).

[36]Joint Command and Control Warfare doctrine is established in Joint Publication (JP) 3-13.1. It defines C2W as, "The integrated use of operations security, military deception, psychological operations, electronic warfare, and physical destruction, mutually supported by intelligence, to deny information to, influence, degrade or destroy adversary command and control capabilities, while protecting friendly command and control capabilities against such actions."

[37]Draft, *Department of Defense IO Roadmap,* Office of the Secretary of Defense, 2003.

[38]Public affairs operations are also referred to media operations.

[39]Thom Shanker and Eric Schmitt, "Firing Leaflets and Electrons, U.S. Wages Information War," *New York Times* (24 February 2003), accessed from http://www.us.army.mil/portal/jhtml/earlyBird/Feb2003/e20030224156370.html 15 Jul 03.

[40]Senior U.S. Defense Official, "Background Briefing on Enemy Denial and Deception" (24 October 2001), accessed October 26, 2001 from http://www.defenselink.mil/news/Oct2001/g011024-D-6570C.html).

[41]Terrence Smith (interviewed by Peter Jennings), "Window on the War," The PBS Online News Hour (8 October 2001), accessed March 19, 2003 from http://www.pbs.org/newshour/bb/media/july-dec01/jazeera_10-8.html.

[42]Jonathan Braude *GlobalSecurity News,* accessed February 2, 2002 from www.globalsecurity.org/military/library/news/2001/09/mil-010921-2a613963.htm.

[43]Associated Press, "U.S. Military Turns to Video of Sept. 11 to Win the Hearts and Minds of Afghans," accessed August 13, 2003 from http://www.whdh.com/news/articles/world/A6392/.

[44]Herbert A. Friedman, "Psychological Operations in Afghanistan," accessed August 13, 2003, from http://www.psywarrior.com/Herbafghan.html.

[45]Ibid.

[46]Associated Press,"Pamphlets Found in Afghanistan Put Price on the Head of Westerners" (6 April 2002), accessed August 13, 2003 from http://www.hollandsentinel.com/stories/040602/new_040602002.shtml.

[47]"Electronic Warfare: Comprehensive Strategy Still Needed for Suppressing Enemy Air Defenses," General Accounting Office report to the Secretary of Defense. GAO–03-51-*Electronic Warfare* (November 2002), 5.

[48]"Operation Iraqi Freedom by the Numbers," USCENTAF (2003), 8.

[49]"U.S. Takes to the Air (waves) in Afghan Campaign," *Christian Science Monitor,* Vol. 93 Issue 231 (October 2001): 7, 24.

[50]Thom Shanker and Eric Schmitt, "Firing Leaflets and Electrons, U.S. Wages Information War," *New York Times* (24 February 2003), accessed July 15, 2003 from https://www.us.army.mil/portal/jhtml/earlyBird/Feb2003/e20030224156370.html.

[51]Ibid.

[52]Douglas Redman and Jack Taylor, "Intelligence and Electronic Warfare Systems Modernization," *Army Magazine* (April 2002), accessed July 15, 2003 from http://www.ausa.org/www/armymag.nsf/0/DCD767B01BDF0AE685256B83 006F1BE7?OpenDocument.

[53]"Operation Enduring Freedom—One Year of Accomplishments," USEUCOM Public Affairs Office (7 October 2002), accessed July 15, 2003 from http://www.eucom.mil/Directorates/ECPA/index.htm?http://www.eucom.mil/ Directorates/ECPA/Operations/oef/operation_enduring_freedomII.htm&2.

[54]Donald Rumsfeld, "Web Site OpSec Discrepancies, Message 141553Z Jan 03," *Defense Link* (14 January 2003), accessed July 9, 2003 from http://www.dod .gov/webmasters/policy/rumsfeld_memo_to_DOD_webmasters.html.

[55]Elizabeth Becker, "In the War on Terrorism, a Battle to Shape Opinions," *New York Times* (11 November 2001): B4.

[56]Chris Hogg, "Embedded Journalism and the Rules of War," *DigitalJournal.Com* (10 April 2003), accessed July 10, 2003 from http://www.digitaljournal.com/ print.htm?id=3586.

[57]Paul Sperry, "Marine General Slams 'Chicken Little' News," *WorldNetDaily.com* (2 July 2003), accessed July 2, 2003, from http://www.worldnetdaily.com/news/ article.asp?ARTICLE_ID=33378.

[58]Bradley Graham, "Bomber Crew Recounts Accounts Attack Operation," *Washington Post,* 8 (April 2003), accessed July 15, 2003.

[59]Andrew Buncombe, "U.S. Army Chief Says Iraqi Troops Took Bribes to Surrender," *The Independent* (London) (24 May 2003), accessed July 15, 2003 from http://news.independent.co.uk/world/middle_east/story.jsp? story=409090.

[60]Kenneth Freed, "U.S. Gave Cedras $1 Million in Exchange for Resignation," *Los Angeles Times* (14 October 1994), accessed July 14, 2003 from http://www-tech .mit.edu/V114/N48/cedras.48w.html.

[61]Michael Kilian, "U.S. Targets Iraqis' Resolve with Psychological Warfare," *Chicago Tribune* (3 February 2003), accessed July 16, 2003 from http:// www.kansascity.com/mld/kansascity/news/5258152.htm?template= contentModules/printstory.jsp.

[62]Ibid.

[63]Defense Science Board, *The Creation and Dissemination of All Forms of Information in Support of Psychological Operations (PSYOPS) in Time of Military Conflict"* (Office of the Undersecretary of Defense for Acquisition, Technologies, and Logistics, Washington, D.C., May 2000), 48.

[64]"HMS Chatham News," UK Ministry of Defense (9 April 2003), accessed July 8, 2003 from http://www.royal-navy.mod.uk/static/pages/content.php3? page=5034.

[65]One of Baghdad Bob's more memorable one-liners.

[66]John Yurechko, "Iraqi Deception and Denial," U.S. State Department Briefing (11 October 2002), accessed July 4, 2003 from http://fpc.state.gov/14337.htm. Note: This briefing to the media was intended to show some of the denial and deception techniques the Iraqis had used.

[67]"Mobile Labs Found in Iraq," CNN.com (15 April 2003), accessed April 15, 2003 from http://us.cnn.com/2003/WORLD/meast/04/14/sprj.irq.labs/.

[68]David Jackson, "U.S. Confident on Weapons," *Dallas Morning News* (16 April 2003), accessed July 11, 2003 from http://www.bradenton.com/mld/charlotte/news/special_packages/iraq/5643233.htm?template=contentModules/printstory.jsp.

[69]John Yurechko, "Iraqi Deception and Denial," U.S. State Department Briefing (11 October 2002), accessed July 4, 2003 from http://fpc.state.gov/14337.htm. Note: This briefing to the media was intended to show some of the denial and deception techniques the Iraqis had used.

[70]"Iraq Lessons Learned: Combat Lessons Learned," *The Strategy Page,* accessed July 14, 2003 from http://www.strategypage.com/iraqlessonslearned/iraqwarlessonslearned.asp.

[71]"Saddam's Surprisingly Friendly Skies," *Business Week Online* (4 April 2003), accessed July 7, 2003 from http://www.businessweek.com/technology/content/apr2003/tc2003044_3053_tc119.htm.

[72]Ibid.

[73]Ibid.

[74]Jim Garamone, "CENTCOM Charts Operation Iraqi Freedom Progress," *American Forces Press Service* (25 March 2003), accessed July 9, 2003 from http://www.defenselink.mil/news/Mar2003/n03252003_200303254.html on 9 Jul 03.

[75]Thom Shanker and Eric Schmitt, "Firing Leaflets and Electrons, U.S. Wages Information War."

[76]"Intercepts: Cyber Battles," *Federal Computer Week* (7 April 2003), accessed April 15, 2003 from http://www.fcw.com/fcw/articles/2003/0407/intercepts-04-07-03.asp.

[77]Dennis Fisher, "Clarke Takes Gov't to Task Over Security," *eWeek* (15 July 2003), accessed July 8, 2003 from http://www.eweek.com/print_article/0,3668,a=44781,00.asp.

[78]General Tommy Franks, "Statement Before the Senate Armed Services Committee" U.S. Senate update (7 February 2002).

[79]J. Michael Waller, "Losing a Battle for Hearts and Minds," *Insight Magazine* (1 April 2002), accessed April 9, 2002 from http://www.insightmag.com/main.cfm/include/detail/storyid/225520.html.

[80]Margaret H. Belknap, "The CNN Effect: Strategic Enabler or Operational Risk?" *Parameters* (Autumn 2002): 110.

[81]J. Michael Waller, op. cit.

[82]Jim Garamone, "General Myers Speaks about the Importance of Focused National Power," *Armed Forces News Service* (16 November 2001), accessed August 13, 2003 from http://www.af.mil/news/Nov2001/n20011116_1644.shtml.

[83]Vice Admiral Thomas R. Wilson, "Testimony Before the Senate Select Committee on Intelligence" U.S. Senate update (6 February 2002).

[84]Jim Garamone, "CENTCOM Charts Operation Iraqi Freedom Progress," *American Forces Press Service* (25 March 2003), accessed July 9, 2003 from http://www. defenselink.mil/news/Mar2003/n03252003_200303254.html on 9 Jul 03.

Chapter 5

[1]"National Security Strategy of the United States." The White House, Washington, D.C., September 2002; pg. 30.

[2]"Fact Sheet: Theater Engagement Planning Review." http://www.dtic.mil/dpmo/pr/factsheet26.htm, 4 Feb 04.

[3]Many of these lessons can be located at the Center for Army Lessons Learned (CALL) website at http://call.army.mil.

[4]The Army's 1st IO Command also supports U.S. Marine Forces in some instances.

[5]Smith, Jeremy D., and Haysman, Peter J. "Using the Strategy to Task Technique to Prioritise Technology Options." www.argospress.com/jbt/Volume5/5-2-6.pdf, 31 Jan 04.

[6]Joint Chiefs of Staff, Joint Publication 1-02, *Department of Defense Dictionary of Military and Associated Terms.* Washington, D.C, 12 April 2001; pg. 284.

[7]A great number of excellent resources exist which can provide an authoritative and exhaustive discussion of these issues. At the least these include the following, from which much of the text of this chapter is developed:

(a) *An Assessment of International Legal Issues in Information Operations,* a paper written by the U.S. Department of Defense Office of General Counsel and released in May 1999.

(b) Michael N. Schmitt, *Computer Network Attack and the Use of Force in International Law: Thoughts on a Normative Framework,* 37 Colum. J. Transnat'l. L. 885 (1999).

(c) Michael A. Sussmann, *The Critical Challenges from International High-Tech and Computer-Related Crime at the Millenium,* 9 Duke J. Comp. & Int'l L. *451* (Spring, 1999).

[8]U.N. Charter, Art. 1.

[9]For an outstanding discussion of this balance and its basis in international jurisprudence, see Col. James P. Terry, USMC (Ret.), *Responding to Attacks on Critical Computer Infrastructure: What Targets? What Rules of Engagement?* 46 Naval L. Rev. 170 (1999).

[10]U.N. Charter. Art. 24.

[11]Compare the text in Articles 41 and 42 to the language in Article 2(4) and Article 51.

[12]U.N. Charter, Art. 48–49.

[13]See, for example, Article 86 of the Law of the Sea Convention and Article 35 of the International Telecommunications Convention.

[14]Article 38, International Telecommunications Convention.

[15]On this aspect, see the discussion on page 4 of the DoD General Counsel Memorandum, cited Note 1, *supra.*

[16]The Law of Armed Conflict recognized the likelihood of applying the protections in the face of technological change. For example, the preamble to the Hague Convention IV declares that, in cases not specifically addressed, civilians and combatants "remain under the protection and the rule of the principles of the laws of nations, as they result from the usages established among civilized peoples, from the laws of humanity, and from the dictates of the public conscience."

[17]See DoD General Counsel Memorandum, *supra* Note 1, at pp. 6–7.

[18]See NWP 1-14M *Commander's Handbook on the Law of Naval Operations,* Section 8.1.1.

[19]Additional Protocol I to the Geneva Conventions, Art. 52(2).

[20]*Supra,* Note 12.

[21]Hague Convention IV, Art. 24.

[22]Unclassified Brief *Information Operations: Legal and ROE Issues* provided by Phillip A. Johnson, consultant to ASD C31 to the Law of Military Operations classes at the Naval Justice School, Newport, Rhode Island.

[23]These include, at the federal level, the following statutes:
    18 USC 1029: Access Device Fraud
    18 USC 1030: Computer Fraud Act
    18 USC 2500 and 2511: Wiretaps and Other Interception and Disclosure of Wire, Oral, or Electronic Communication
    18 USC 2701: Stored Wire and Communication Access
    18 USC 1343: Wire Fraud
    18 USC 1363: Malicious Mischief
    18 USC 1367: Interference with Satellites
    18 USC 2701: Stored Electronic Communications

[24]Michael A. Sussman, *The Critical Challenge.*

[25]Ibid, p. 471.

Chapter 6

[1]Jen Jui-Wen, "Latest Trends in China's Military Revolution," translation from *FBIS*-CHI-96-047, pp. 1–4.

[2]Russian National Security Concept, Nezvisimoye Voyennoye Obozreniye, 17 December 1999, accessed from http://www.acronym.org.uk/dd43/43NSC.htm on 23 January 2004.

[3]E. A. Belaev, "Informatsionnaya bezopasnost' kak global'naya problems" [Information Security as a Global Problem], a chapter in *Global'nye problemy kak istochnik chrezvychaynykh situatsiy* [Global Problems as a Source of Emergency Situations] (location: URSS, 1998), edited by Iu. L. Vorob'ev, p. 125.

[4]Joseph Nye, "Redefining the National Interest," *Foreign Affairs,* July/August 1999. Vol. 78, No. 4, p. 26.

[5]Ibid, p. 24.

[6]Ibid.

[7] A. A. Kokoshin, "Voenno-politicheskie i ekonomicheskie aspekty reformy vooruzhennykh sil Rossii [Military-Political and Economic Aspects of Reform of the Russian Armed Forces], *Voyennaya Mysl* [*Military Thought*] No. 6, (1996): 9.

[8] V. I. Tsymbal, "Kontseptsiya 'informatsionnoy voyny' " [Concept of Information Warfare], talk given at a conference in Moscow in September 1995, p. 7.

[9] *"The Russian View of Information War,"* by Timothy L. Thomas. Foreign, military studies office, Fort Leavenworth, KS accessed at http://fmso.leavenworth.army .mil/FMSOPUBS/ISSUES/Russianview.htm on 23 January 2004.

[10] S. P. Rastorguev, *Informatsionnaya Voina* [Information War] (Moscow: Radio and Communication, 1998).

[11] V. D. Tsigankov and V. N. Lopatin, *Psikhotronnoe oruzhie i bezopasnost' rossii* [Psychotronic Weapons and the Security of Russia] (Moscow: Sinteg, 1999).

[12] Even the military has written about the subject of psychotronic weapons in its publications. For example, see I. Chernishev, "Polychat li poveliteli 'zombi' blast' nad mirom," [Can a Ruler Make 'Zombies' Out of the World]," *Orientir* [Orienteer] (February 1997): 58–62.

[13] Marshal Igor Sergeyev, comments in *Kraznaiya Zvesda* [*Red Star*] (9 December 1999) accessed at http://fmso.leavenworth.army.mil/FMSOPUBS/ISSUES/ Russianview.htm on 23 January 2004. All of Sergeyev's comments in the next eight paragraphs are from this document.

[14] Based on a discussion with modelers at the General Staff Academy in December, 1991.

[15] Author's discussion with General-Major (retired) V. D. Riabchuk, Fort Leavenworth, September 1996.

[16] Ibid.

[17] "NATO's Role in Kosovo." http://www.nato.int/kosovo/kosovo.htm, 1 February 2004.

[18] Hubbard, Zachary P. "Information Operations and Information Warfare in Kosovo: A Report Card We Didn't Want to Bring Home." Cyber Sword, Spring 2000, vol. 4, no. I. pp. 27–29.

[19] Giulio Douhet's described his airpower theories in his book, *Command of the Air,* published in 1921.

[20] David A. Fulghum. "Pentagon Dissecting Kosovo Combat Data," *Aviation Week and Space Technology,* 26 July 1999, pp. 68.

[21] "Apache Helicopter Crashes in Albania." http://www.aircav.com/kosovo/ ah64crash/ah64crash.html, 2 February 2004.

[22] Arkin, William and Windrem, Robert. "The Other Kosovo War." http://seclists .org/lists/isn/2001/Aug/0193.html, 28 January 2004.

[23] Lt. Gen. Huai Guomo, "On Meeting the Challenge of the New Military Revolution," translation from *FBIS*-CHI-96-130, pp. 1–7.

[24] Su Enze, "Logical Concept of Information Warfare,"translation from *FBIS*-CHI-96-135, pp. 1–5.

25Lei Zhoumin, "Information Warfare and Training of Skilled Commanders," translation from *FBIS*-CHI-96-036, pp. 1–5.

26Renzhao, op. cit., pp. 4–5.

27*Ibid*, p. 4.

28Xu Chuangjie, "Military Revolution Gives Impetus to Evolution in Command," translation from *FBIS*-CHI-96-030, p. 1.

29*Ibid.*

30*Ibid*, pp. 1–2.

31Wei Jincheng, op. cit., p. 3.

32Zhou Li and Bai Lihong, "Information Warfare Poses Problems," translation from *FBIS*-CHI-96-014, pp. 1–2.

33Ibid.

34Chou Hsi, "Exploration and Analysis of Military Computer Security and Virus Protection," translation from *FBIS*-CHI-96-116, pp. 1–6.

35Interview between Martin Libicki and Dr. Ahari. The Chinese have proven themselves remarkable in indigenizing Marxism to suit their cultural requirements. They are likely to develop information-based warfare techniques to suit their special needs before too long. The United States must remain especially sensitive to this profound historical reality about the PRC.

36On this, see the government information paper on "Protection of Australia's National Information Infrastructure and E-Security Policy: Administrative and Operational Arrangements—September 2001," available at http://www. noie.gov.au/Projects/information_economy/e-security/nat_agenda.htm; I. Dudgeon, "Protection of the National Information Infrastructure," *Journal of the Royal United Services Institute of Australia,* Vol. 20 (December 1999): 53–58; Interdeparmental Committee on Protection of the National Information Infrastructure, *Protecting Australia's National Information Infrastructure* (Canberra: Attorney General's Department, December 1998). Dudgeon wrote an initial assessment report on the vulnerabilities of the Australian NII in 1996 for the Defence Signals Directorate (DSD). A sanitized version of Dudgeon's report is attached to the above report. It is available via the Internet at: http://www.ag.gov.au/publications/niireport/niirpt.pdf. For a differing (though somewhat dated) view, see A. Cobb, *Australia's Vulnerability to Information Attack: Towards a National Information Policy* (Canberra: Australian National University, Strategic and Defence Studies Centre, Working Paper No. 310, 1997); and the same author's *Thinking about the Unthinkable: Australian Vulnerabilities to High-Tech Risks* (Canberra: Department of the Parliamentary Library, 1997–1998, Research Paper No. 18, 1998). This later report is available via the Internet at http://www.aph.gov.au/library/pubs/rp/ 1997-98/98rp18.htm.

37In this regard, the work of Dr Carlo Kopp on IO and related issues should be noted. On this, see C. Kopp, "Supporting Jamming and Force Structures," *Journal of Electronic Defense* (May 2002): 44–48; "A Fundamental Paradigm of Infowar," *Systems* (February 2000): 47–55; "Issues in Current Infowar,"

Systems (March 2000): 31–38; *Command of the Electromagnetic Spectrum: An Electronic Combat Doctrine for the RAAF* (Canberra: Air Power Studies Centre, Working Paper No. 8, 1992); *An Introduction to the Technical and Operational Aspects of the Electromagnetic Pulse Bomb* (Canberra: Air Power Studies Centre, Working Paper No. 50, 1996). Many of Dr Kopp's publications are available from his website, http://www.csse.monash.edu.au/~carlo/mpubs.html.

[38] M. J. Davies, "The Impact of the Revolution in Military Affairs [RMA] on the Australian Defence Forces [ADF]," *Australian Defence Force Journal,* No. 128 (January/February 1998): 42.

[39] Ibid, pp. 42–43.

[40] P. G. Nicholson, "Controlling Australia's Information Environment or Decision Superiority and War-Fighting," in S. Clarke (ed.), *Testing the Limits: Proceedings of a Conference held by the Royal Australian Air Force in Canberra, March 1998* (Canberra: Air Power Studies Centre, 1998): 155–157.

[41] Ibid., pp. 151–152.

[42] C. A. Jamieson, "Psyops Beyond 2000: Coordinating the Message," *Australian Defence Force Journal,* No. 125 (July/August 1997): 43; G. Peterson, "Psyops and Somalia—Spreading the Good News," *Australian Defence Force Journal,* No. 104 (January/February 1994): 39–40.

[43] M. Evans, *Australia and the Revolution in Military Affairs* (Canberra, ACT: Land Warfare Studies Centre, Working Paper No. 115, August 2001): 8–9. DSTO is broadly the equivalent of DARPA. The paper is available electronically from http://www.defence.gov.au/army/LWSC/Publications/wp% 20115.pdf.

[44] Department of Defence, *Australia's Strategic Policy* (Canberra: Directorate of Publishing and Visual Communications, Department of Defence, December 1997).

[45] Ibid, p. 57.

[46] P. G. Nicholson in S. Clarke (ed.), p. 145. On this, see also P. Dibb, *The Relevance of the Knowledge Edge* (Canberra: Strategic and Defence Studies Centre, Working Paper No. 329, December 1998). From 2000 to 2001, the then Air Vice Marshal Nicholson was the chief knowledge officer (CKO) of the ADO. In 2001, the CKO position was renamed "head knowledge systems" (HKS). This position, and its supporting staff—Knowledge Systems Division (KSD) retains responsibility for IO capability development.

[47] HQAST, *Decisive Manoeuvre: Australian Warfighting Concepts to Guide Campaign Planning,* Interim Edition (Canberra: Directorate of Publishing and Visual Communications, Department of Defence, January 1998). HQAST was raised in January 1997, and is the main operational level headquarters for the ADF. On this, see J. M. Connolly, "Theatre Operations in the ADF," *Journal of the Royal United Services Institute of Australia,* Vol. 19 (December 1998): 73–77.

[48] *Decisive Manoeuvre,* p. 4–1.

[49]B. Alsop, "What are Information Operations? Why should I take any notice?," *Australian Defence Force Journal,* No. 140 (January-February 2000): 31. The ADF's current approach to IO will be dealt with in the next section. ADFWC provides training for the ADF on joint operations, and has the responsibility for the development of joint doctrine. In broad terms it performs a comparable role to the JFSC.

[50]Department of Defence, *Defence: Our Priorities* (Canberra: Defence Publishing Service, November 1998).

[51]*1999 Defence Communications Development Seminar: Shaping the ADF Knowledge Edge—C3I Opportunities and Challenges . . .* Canberra, March 9–11.

[52]P. Mitchell, *Knowledge Operations* (Presentation given at the 1999 Defence Communications Development Seminar). At the time of the presentation, Commodore Mitchell was the Director-General of Information Strategic Concepts (ISC) branch in Strategic Command Division (SCD) of Australian Defence Headquarters (ADHQ). Between 1998 and 1999, ISC branch was responsible for the development of information-related warfighting concepts that would underpin the current and future ADF. In late 1999, as the result of a reorganization of the Defence organization, the ISC Branch was moved from SCD to become part of the Command, Control, Communication, Computers, Intelligence, Surveillance, Reconnaissance and Electronic warfare C4ISREW Capability staff. In 2000, as part of a further refinement of information-related capability responsibilities in the Defence organization, the C4ISREW capability staff was renamed the Knowledge Staff. ISC branch was renamed ISREW branch, and accepted further responsibilities relating to ISR and EW capability development.

[53]Ibid, slides 11–15. See also P. G. Nicholson in S. Clarke (ed.), op. cit., pp. 146–147.

[54]P. Mitchell, op. cit., slide 15. For other views on this, see E. F. Murphy et al., "Information Operations: Wisdom Warfare for 2025," in *Air Force 2025 White Papers,* Vol. I, (Maxwell Air Force Base, Alabama: Air University Press, 1996); and P. Baumard, "From InfoWar to Knowledge Warfare: Preparing for the Paradigm Shift," in A. D. Campden et al. (eds.), *Cyberwar: Security, Strategy and Conflict in the Information Age* (Fairfax, Virginia: AFCEA International Press, 1996), 147–160.

[55]In particular, the work of DSTO scientist Martin Burke is noteworthy here. See his *Thinking Together: New Forms of Thought Systems for a Revolution in Military Affairs* (DSTO-RR-0173 dated July 2000); *Information Superiority, Network Centric Warfare and the Knowledge Edge* (DSTO-TR-0997 dated July 2000); and *Thought Systems and Network Centric Warfare* (DSTO-RR-0177 dated July 2000). These documents are available electronically from http://www.dsto.defence.gov.au/corporate/reports/index.html.

[56]Department of Defence, *Defence 2000—Our Future Defence Force* (Canberra, ACT: Defence Publishing Service, 2000). This document is available electronically from http://www.defence.gov.au/whiTSCPaper/ (henceforth referred to as *White Paper 2000*); and *Department of Defence, Defence Capability Plan*

2001–2010—*Public Version* (Canberra, ACT: Defence Publishing Service, 2001). This document is available electronically from http://www.defence. gov.au/dmo/id/dcp/dcp_public.pdf (henceforth referred to as *Defence Capability Plan*).

[57]*White Paper 2000*, pp. 94–97.

[58]A. Gibbs, "Integrating IO and Force Level EW," Presentation delivered to the Association of Old Crows 4th Australian National Convention, Canberra, 19 February 2002.

[59]*RAN Doctrine 1—Australian Maritime Doctrine* (Canberra, ACT: RAN Sea Power Centre, 2000). This document is available electronically from http://www.navy.gov.au/seapowercenter/maritimedoctrine.htm, *Land Warfare Doctrine 1—The Fundamentals of Land Warfare* (Puckapunyal, VIC: Land Warfare Development Centre, 2002), and *Fundamentals of Australian Aerospace Power* (Canberra, ACT: Aerospace Centre, 2002).

[60]ADDPs articulate joint doctrine for the ADF. They are the equivalent of Joint Publications (JP) in the U.S. military system.

[61]The following paragraphs are derived from the chronology provided in Joint Standing Committee on Foreign Affairs, Defence and Trade, *Bougainville: The Peace Process and Beyond* (Canberra: Australian Government Publishing Service, September 1999), Appendix D—Outline History of the Bougainville Conflict, pp. 169–185. This report is available via the Committee website on the Internet at http://www.aph.gov.au/house/committee/jfadt/bougainville/BVrepindx.htm. Another chronology of the Bougainville Conflict is available from the Department of Foreign Affairs and Trade (DFAT) website at http://www.dfat.gov.au/geo/png/bougainville/png_bg.html.

[62]On Operation LAGOON, see D. Grevis-Jones, "Bougainville—Island of Agony," *Army Magazine*, No. 21 (1994): 30–33; M. Miko, "We gave peace a chance—Operation LAGOON, Bougainville 1994," *Naval Supply Newsletter: Journal of the Royal Australian Navy Supply Support Organisation*, Vol. 5, No. 6 (June 1995): 43–46; B. Breen, *Giving Peace a Chance—Operation LAGOON, Bougainville 1994: A Case of Military Action and Diplomacy* (Canberra, ACT: Strategic and Defence Studies Centre, Canberra Paper No. 142, 2001).

[63]On the Sandline Affair, see S. Dorney, *The Sandline Affair: Politics and Mercenaries and the Bougainville Crisis* (Sydney: ABC Books, 1998); M. O'Callaghan, *Enemies Within: Papua New Guinea, Australia and the Sandline Crisis—The Inside Story* (Sydney: Doubleday, 1999).

[64]The text of the Burnham Declaration is contained in the Joint Standing Committee on Foreign Affairs, Defence and Trade, op. cit., pp. 187–189.

[65]The text of the Burnham Truce is contained in ibid, pp. 191–196.

[66]Ibid, p. 72.

[67]The text of the Lincoln Agreement is contained in ibid, pp. 197–201. The text of the Arawa Agreement is also contained here, pp. 203–209.

[68]P. Clark, "The Road to Peace: Aspects of Information Operations Applied to Peace Monitoring Operations on Operation BELISI, Bougainville 1997–98," *Combat Arms*, 2/98, p. 61.

[69]P. Clark, "MIST in Bougainville: Military Information Support to Peace Operations," in G. Fry and M. McDonald (eds.), *1998—The Australian Army in Profile* (Canberra: Defence Public Affairs Organisation, November 1998), 86.

[70]P. Clark, "The Road to Peace," p. 67.

[71]P. Clark, "MIST in Bougainville," p. 87.

[72]P. Clark, "The Road to Peace," p. 68. *Hombru* production and consumption was targeted by the MIST, as this was identified as a key factor in sparking incidences of violence between the factions in Bougainville.

[73]The following passages are derived from the chronology provided at the U.N. Internet site: http://www.un.org/peace/etimor99/chrono/chrono_frame.html. For a general overview of the history of East Timor and Australia, see J. Cotton (ed.), *East Timor and Australia* (Canberra: Australian Defence Studies Centre, Australian Defence Force Academy, 1999). See also D. Greenlees and R. Garran, *Deliverance: The Inside Story of East Timor's Fight for Freedom* (Crow's Nest, NSW: Allen & Unwin, 2002). For an official Australian account, see Department of Foreign Affairs and Trade, *East Timor in Transition 1998–2000: An Australian Policy Challenge* (Canberra, ACT: Department of Foreign Affairs and Trade, 2001).

[74]For a general overview of Operation STABILISE, see B. Breen, *Mission Accomplished, East Timor: The Australian Defence Force Participation in the International Force East Timor* (St Leonards, NSW: Allen & Unwin, 2001); A. Ryan, *Primary Responsibilities and Primary Risks: Australian Defence Force Participation in the International Force East Timor* (Canberra, ACT: Land Warfare Studies Centre, Study Paper No. 304, November 2000). Ryan's work is available electronically at http://www.defence.gov.au/army/lwsc/Publications/SP%20304.pdf.

[75]A soon-to-be-released monograph deals with the conduct of IO during Operation STABILISE in detail: K. Beasley, *Information Operations during Operation STABILISE in East Timor* (Canberra, ACT: Land Warfare Studies Centre, Working Paper No. 120, August 2002). Presently available works include J. Blaxland, "On Operations in East Timor: The Experiences of the Intelligence Officer, 3rd Brigade," *Australian Army Journal,* Issue 2000, pp. 1–12; and the same author's *Information-Era Manoeuvre: The Australian-Led Mission to East Timor* (Canberra, ACT: Land Warfare Studies Centre, Working Paper No. 118, June 2002). The working paper is available electronically from http://www.defence.gov.au/army/lwsc/Publications/WP%20118.pdf.

[76]In July 2000, Cosgrove was promoted to the rank of lieutenant general and appointed Chief of Army. In June 2002, he was promoted to general and appointed Chief of the Defence Force. Transcript of Address by (then) Major General P. G. Cosgrove . . . *The ANZAC Lecture,* Georgetown University (4 April 2000), 7.

[77]Ibid.

[78]On this, see R. H. Scales, "Trust, Not Technology, Sustains Coalitions," *Parameters* (Winter 1998–99): 9; R. Brennan and R. Evan Ellis, *Information Warfare in Multilateral Peace Operations—A Case Study of Somalia* (Prepared for the Office of the Secretary of Defense, Net Assessment by SAIC, Project No. 03-9847-000, 18 April 1996), ii.

[79]In this regard, see C. de Caro, "Softwar," in A. D. Campden et al. (eds.), op. cit., pp. 208–209.

[80]P. G. Cosgrove, op. cit., p. 7.

[81]J. Martinkus, "Cosgrove Rejects Infiltration Claims as Simply Rubbish," *Canberra Times* (15 October 1999): 2.

[82]P. G. Cosgrove, op. cit., p. 8. In this regard, see also A. E. Dowse, "Harnessing the News Media During Conflict," *Australian Defence Force Journal*, No. 129 (March/April 1998): 43–51.

[83]On this matter, see M. Smith, "Analysis of the NATO Information Campaign—in the War Against Yugoslavia," *Australian Defence Force Journal*, No. 140 (January/February 2000): 49–54.

[84]For one example, see A. Ryan, *From Desert Storm to East Timor: Australia, the Asia-Pacific and the "New Age" Coalition Operations* (Canberra: Land Warfare Studies Centre, Study Paper No. 302, January 2000). The work is available electronically at http://www.defence.gov.au/army/lwsc/Publications/sp%20302.pdf.

About the Contributors

Ehsan M. Ahrari, Professor of National Security and Strategy, Joint Forces Staff college. Dr. Ahrari contributed the section on Chinese IW efforts, as well as provided guidance for strategic IO interests. He has served as part of the senior staff at the Joint Forces Staff College since 1994, working to build the IO curriculum for the Joint Command Warfare School. His previous teaching assignments include the United States Air Force Air War College, Mississippi State University and East Carolina University. A graduate of Southern Illinois University (1976), Dr. Ahrari is published widely with two books and numerous scholarly articles to his credit.

Leigh Armistead, Civil Servant, US Navy. Serving as the primary editor for this book. He is a retired naval officer with a number of staff and operational tours in Airborne Early Warning (AEW) and C3 units, including E-2C Hawkeye squadrons, USS *Nimitz* (CVN-68), Marine Action Weapon and Tactics squadron One (MAWTS-1),Task Force Web And Commander US Naval Forces Central Command (COMUSNAVCENT) Formerly an IW instructor at Joint Forces Staff College, he is currently an information assurance officer at Fleet Forces Command. A graduate of the U.S. Naval Academy, U.S. Navy, and U.S. Air Force Command and Staff College, Leigh Armistead is finishing a Ph.D. program at Edith Cowan University, Perth, Australia, where he is writing his dissertation on IO. He has written two books and a number of articles for professional journals, as well as hosted numerous professional conferences on IO.

Major Robert E. Blackington, United States Air Force. Currently the Chief of Initiatives at the Space Warfare Center, Major Blackington was previously a student at the Air Command and Staff College at Maxwell AFB, and before that an instructor at the Joint Forces Staff College. He spent the majority of his career flying the MC-130E and AC-130E gun ship aircraft as a navigator and has over 3,000 hours in type. While serving as an IW instructor at JFSC, he was instrumental in upgrading the U.S. Air Force IO community's awareness of the role of the former USIA as well as IPI and PDD-68. In addition to providing guidance in these areas, Major Blackington was the main contributor for the section on EW.

Byron Collie, formerly a Federal Agent with Australian Federal Police Technical Operations, Australia. Byron Collie is an experienced computer crime investigator, well known in the international computer security incident response community, and has worked a number of international computer intrusion cases with U.S. military and federal law enforcement agencies including *Solar Sunrise*. From 1998 to 2000, Mr. Collie was attached to the Royal Australian Air Force (RAAF) Directorate of IW where he was a Project Manager. During that period he both graduated from the 1998 Australian Defence Force Warfare Centre inaugural Information Operations Staff Officers Course (IOSOC) and instructed in IA and CND in that year, and in the 1999 IOSOC. Mr. Collie is currently employed by the Wells Fargo Bank in Minneapolis as a Senior Information Security Engineer.

Captain Carlton T. Fox, Jr., U.S. Air Force Intelligence Officer. A 1999 graduate of the Virginia Polytechnic Institute and State University, Captain Fox earned his bachelor's degree in Political Science and a master's degree in History with a concentration in Foreign Relations. A key ingredient in the overall success of this book, he served as the assistant editor for this project in edition to writing the section on counterterrorism. An essential player in bringing this book to print on a timely basis, Captain Fox is currently fighting the Global War against Terrorism in the joint environment at Fort Hood Army Installation, Texas.

Mark R. Goodell, Staff Officer, U.S. Space Command. Lieutenant Colonel Goodell is currently attached to SpaceCom, having previously served in a four-year tour as a C4I Instructor at JFSC. A career Space Controller, Lt. Col. Goodell was stationed in a variety of positions from Flight Commander at Falcon AFB, to Atlas II Launch Controller, Crew Commander at the 73rd and 16th space Surveillance Squadrons. In addition, he also served as a Staff Officer, HQ USAF Space Command. A Graduate of the Air Force Institute of Technology and Brigham Young University, Lt. Col. Goodell has also completed the Air Command and Staff College and Squadron Officer School, and contributed the section on space and IO for this book.

D ave Harris, Lieutenant Colonel (LTC) Australian Regular Army (Ret.) Dave Harris served for twenty-five years in a variety of billets in the Royal Australian Armoured Corps, including tank gunner/operator and regimental appointments in tank, reconnaissance, and armoured personnel carrier regiments. He has also served with the Royal Canadian Dragoons and the United Nations (UNIIMOG), as well as instructing tactics and leadership at the Royal Military College, Duntroon. His last posting in the Australian Defence Force was as the Information Operations Planner in the Directorate of Joint Plans in Strategic Command.

Z achary P. Hubbard, Lieutenant Colonel (LTC) U.S. Army (Ret.) Former chief of the Information Warfare Division, Joint Forces Staff College from April 1998–April 2001, (LTC)Hubbard was a prime advocate for the publication of this book. Commissioned in the Field Artillery, he is also qualified as a counterintelligence and human intelligence officer. (LTC) Hubbard's operational experience includes service in operations Desert Shield and Desert Storm, Saudi Arabia/Kuwait; JTF Andrew, Florida; CJTF Kismayo, Somalia; Operations Sharp Guard and Deny Flight, Italy; and IFOR/SFOR in Bosnia-Herzegovina. He is currently working for Zeltech Corporation in Hampton, Virginia, as an Information Assurance Analyst, supporting the US Air Force at Langley AFB.

R ichard J. Kilroy, Jr., Information Warfare Instructor, Assistant Professor, Joint Forces Staff College. Army Intelligence Officer who has served in a variety of Tactical and Strategic Intelligence Staff Officer positions in the United States and in Europe. LTC Kilroy (Ret.) was a Latin American Foreign Area Officer, who served in a variety of politico-military affairs positions in the U.S. Southern command, including serving as a Special Assistant to General Barry McCaffrey and General Wes Clark. He attended the Mexican Command and General Staff College and has authored articles on civil-military relations in Latin America. Rick Kilroy holds an M.A. and Ph.D. in Foreign Affairs from the University of Virginia.

D an Kuehl, Professor Information Resources Management College (IRMC) at National Defence University (NDU). Dr Kuehl is the Director of the Information Strategies Concentration Program; a specialized curriculum for selected students at the NWC and ICAF. Lieutenant

Colonel Kuehl (ret.—USAF) served primarily as a Minuteman ICBM Instructor Crew Commander, Nuclear Planner at HQ SAC, and on the Air Staff during Operation Desert Shield/Storm. He supported the landmark Gulf War Air Power Survey, and authored the "Air Campaign" chapter in the DoD'S *Final Report to Congress on the Persian Gulf War.* He has numerous articles published in journals contributing to the IO and EW fields, and has also co-edited the pending book *Cyberwar 4.0: Information Operations: Applying Power in the Information Age.*

Jeff Malone, Australian Regular Army. A prior enlisted soldier, Jeff Malone was commissioned in 1992, and has served in a variety of regimental and staff appointments, including appointments in the information operations area. Captain Malone is currently posted to the Directorate of Strategy, Future Land Warfare Branch, Army Headquarters. He has a B.A. (Honours) and an M.A. (Research) in Political Science from the University of Western Australia, Perth. At present, he is completing a Ph.D. dissertation entitled "Information Operations and Australian National Security Policy" at the Australian Defence Force Academy in Canberra. In September 2002, Captain Malone won the Chief of Defence Force Scholarship to continue work on his Ph.D. in 2003.

Robert J. Orr, Commander (CDR), U.S. Navy, Staff Judge Advocate for Commander, Navy Region Mid-Atlantic. Previously Robert J. Orr served as Fleet Judge Advocate Commander in Chief, U.S. Atlantic Fleet. A graduate of Heidelberg College in 1981, CDR Orr received his Juris Doctor from Ohio State University in 1984. He also holds a Masters of Law in International Law from the University of Virginia and is a graduate of the College of Command and Staff, U.S. Naval War College. CDR Orr has a strong operational background, with operational cruises on the USS *Midway* (CV-41), USS *Eisenhower* (CVN-69), and USS *Saratoga* (CV-60). CDR Orr is an adjunct instructor for the Joint Forces Staff College and the U.S. Naval War College.

Neil Quarmby, Lieutenant Colonel (LTC) Australian Regular Army. Commissioned into the Australian Intelligence Corps from the royal Military College in 1984, Neil Quarmby served regimental appointments in Australia's medium artillery regiment and electronic warfare regiment. He has served with the British Army on the Rhine and has been involved in a number of counterintelligence and counterterrorist operations and duties. With masters degrees in International Relations and Defence Studies, he has been active in Australian defence capability development and is currently serving in the Defence Intelligence Organisation. LTC

Quarmby contributed to a number of sections throughout the book and has been the driving force behind IO training and doctrine in the Australian Defence Force.

Tim Thomas, Foreign Military Studies Office, Ft. Leavenworth, Lieutenant Colonel (LTC) Thomas, U.S. Army (Ret.) is a guest speaker for the JFSC JIWSOC and JIWOC sessions as well as a nationally recognized expert on Russian and Chinese IW doctrine. He was the featured speaker at the Information Warfare Convention 2000 in Washington, D.C., and contributed mostly to the Russian IW section of this text.

Index

Note: Page numbers in *italics* indicate diagrams.

For a list of abbreviations and acronyms please see pages vii-xiv.

Abshire, David, 136
Academic theory, 14–15
"Access to Society: A Neglected Dimension of Power" (Haskell), 11
ADDP. *See* Australian defence doctrine publication
ADF. *See* Australian Defence Force
ADHQ. *See* Australian Defence Headquarters
Adversaries.
Afghanistan, 93, 146–47
  post-Taliban and, 151–52
  war with Soviet Union, 14
  See also Operation Enduring Freedom (OEF)
AFIWC. *See* Air Force Information Warfare Center
Aideed, first name ???????, 16
Air Force Information Warfare Center (AFIWC), 23, 166
Al Barakaat, 143
Al Jazeera, 148, *155*
al Qaeda, 4, 92–93, 137, 141–42, 144, 158
  blocked financing, 143
Al Taqwa, 143
Al-Sahhaf, Saaed ("Baghdad Bob"), 155
Alfred P. Murrah Federal Building, 91
  See also Oklahoma City bombing
Anthrax scare, 161
Anti-terrorism, 92
  See also Counterterrorism
Anti-U.S. attacks (1996–1999), *105*
Apaches, 203
Arquilla, John and David Ronfeldt
  Emergence of Noopoltik: Toward an American Information Strategy, 9, 15
  In Athena's Camp, 11
Article V (NATO), 143
Asymmetric non-nuclear attacks, 21
Asymmetric threats, 14
Attack plan generators, 86
Australian Army, 218

Australian defence doctrine publication (ADDP), 218
Australian Defence Force (ADF), 212–29
  Bougainville (Papua New Guinea) conflict, 219–24, 228–29
    background, 219–22
    IO contributions to Operation BELISI, 222–24
    map, *219*
    structure of the Peace Monitoring Group, *223*
  decision superiority and subordinate concepts, *216*
  decisive maneuver concept, *215*
  doctrinal approach to IO, 217–18
  East Timor peace-enforcement mission, 224–29
    background, 224–25
    IO contributions to Operation STABILISE, 225–28
    maps, *224, 226*
  evolution of IO and related concepts, 212–17
  unclassified ADF IO doctrine, *217*
Australian Defence Headquarters (ADHQ), 213
Australia's Strategic Policy (ASP97), 214
Aviation Week and Space Technology, 203
Axis of Evil, 134

"Baghdad Bob" (Saaed Al-Sahhaf), 155
Battlespace. *See* Intelligence support
Becker, Elizabeth, 153
Belaev, E. A., 192
Bering, Helle, 131
BIG. *See* Bougainville Interim Government
Bilateral IO Steering and Working Groups (BIOSG/BIOWG), 40
Bill of Rights, 5
bin Laden, Osama, 3–4, 138, 141, 144
  Al Jazeera tapes, 148
Biological agent production labs, Karbala, Iraq, 156
Biometrics, 87–88
BIOSG/BIOWG. *See* Bilateral IO Steering and Working Groups
Black propaganda, 159–60

265

Blair, Tony, 134, 156
Blue Ribbon Commission on National
Security, 44–45
Bluto, 231
Bogdanov, S., 111
Bougainville Interim Government (BIG), 220
Bougainville (Papua New Guinea) conflict.
See Australian Defence Force
Bougainville Reconciliation Government
(BRG), 222
Bougainville Revolutionary Army
(BRA), 220
Bound to Lead (Nye), 11
BRA. See Bougainville Revolutionary Army
Burnham Truce, 221
Bush administration
critical infrastructure protection (CIP)
policy, 99–102
Department of Homeland Security
(Executive Order 13228), 44–45
international public information (IPI)
and, 133–37

C2W. See Command and control warfare
C4. See Command, control,
communications, and computers
CA. See Civil affairs
Cable News Network (CNN), 5
Cai Renzhao, 209
Campbell, Alistair, 134
Capitol Building, 144
Carter administration, Iranian hostage
crisis, 92
CBS radio, 5
CC. See Combatant commanders
Cedras, Raul, 154
CentCom. See Central Command
Central Command (CentCom), 37, 157
Central Intelligence Agency (CIA), 24–25,
27, 54, 55
CERT. See Computer emergency response
team
Chairman of the Joint Chiefs of Staff
Memorandum of Policy 30 (CJCS
MOP 30), 22–23
Chan, 220
Cheney, Dick, 66
Chiapas, Mexico, 6–7
China and information warfare (IW),
207–12
Chinese IO as a warfighting network,
209–11
future of IO in China, 211–12
CHMP. See Coalition Multi-Level Hexagon
Prototype
CI. See Counterintelligence
CICs. See Communication information
centers
CIP. See Critical Infrastructure Protection

CITAC. See Computer Investigations and
Infrastructure Threat Assessment
Center
Civil affairs (CA), 20
Civil liberties, 5
CIWG. See Critical Infrastructure
Working Group
CJCS MOP 30. See Chairman of the Joint
Chiefs of Staff Memorandum
of Policy 30
Clarke, Richard, 96, 99, 109, 135, 158
Classified materials, 60–63
See also Intelligence support
Clinton administration
Blue Ribbon Commission on National
Security, 44–45
critical infrastructure protection (CIP)
policy, 94–98, 107–9
interagency IO organizations created
by, 45
international public information (IPI)
and, 125–27, 129–31
CNA. See Computer network attack
CNN. See Cable News Network
CNO. See Computer network operations
Coalition Multi-Level Hexagon Prototype
(CHMP), 62–63
Coalition operations, 61–63
Cold war
bipolar cold war era, 2, 21
power in, 13
Collaborative white boarding (CWB), 54
Combat support agencies, 32
Combatant commanders (CC), 27, 36–39
coordinating IO with agencies, 164
NGOs and, 46–47
peacetime engagement and, 163
types, 27, 37–38
Combating Terrorism (PDD-62), 96
Command, control, communications, and
computers (C4), 52, 214
Command and control warfare (C2W),
22–23, 22, 145
in Australia, 213, 214, 217
Command and Control Warfare (CJCS
MOP 30), 22–23
Commando Solo planes, 151, 154–55,
172, 207
Common access card (CAC)
technology, 88
Communication information centers
(CICs), 134
Computer emergency response team
(CERT), 32–33, 166
agencies and architecture, 78
DoD coordination with, 81
Computer Investigations and Infrastructure
Threat Assessment Center
(CITAC), 95

Computer network attacks (CNAs), 20
  defense of, 75–79
  exercises, 24
  legal issues, 115
  offensive, 114–18, *114*
  technical issues, 115–16
  See also Information projection;
    Information protection
Computer network defense (CND), 158
  See also Information protection
Computer network operations (CNO),
    41–42
Computer warfare, 18
*Concept for Future Joint Operations:
  Expanding Joint Vision 2010,* 165
Cosmo, 9
Counterterrorism, 90–110
  anti-terrorism, 92
  Counter-Terrorism (PDD-62), 42, 94–98
  counterterrorism information operations
    (CTIO) target groups, 93–94
  Critical Infrastructure Protection
    (PDD-63), 107–9
    budgets by department (fiscal years
      1998–2000), *110*
    goals of, 108–9
    lead agencies by sector, *108*
  defined, 91–92
  domestic counterterrorism operations,
    102–4
  domestic terrorism, 91
  expenditures for (1996–2001), *103*
  FBI funding for (1995–1999), *103*
  international cooperation, 105–7
  international terrorism, 92
  IO and, 90–91
  partnerships for: "Track Two Diplomacy,"
    104–7
  policy tenets, 93
  post-September 11th CTIO
    organizational structure, *102*
  pre-September 11th CTIO organizational
    structure, *100*
  terrorism defined, 91–92
  terrorist incidents in the U.S.
    (1993–1999), *97*
  total anti-U.S. attacks (1996–1999), *105*
  See also Information projection;
    Information protection; Terrorism
Counter-Terrorism and National
  Preparedness Policy Coordination
  Committee, 99
Counter-Terrorism (PDD-62), 42, 94–98
Counter-Terrorism Security Group, 97
Counterintelligence (CI), 52
Counterterrorism information operations
  (CTIO), 91, 93–94
  post-September 11th organizational
    architecture, *102*

pre-September 11th organization
    structure, *100*
  terror groups, 93–94
  See also Counterterrorism
Criminal terrorism, 94
"Critical Challenges from International
  High-Tech and Computer-Related
  Crime at the Millennium"
  (Sussman), 184
Critical Foundations (Commerce
  Department), 42–43, 95, 108
Critical Infrastructure Coordination Group,
    96, 97
Critical infrastructure protection (CIP) policy
  Bush administration, 99–102
  Clinton administration, 94–98
  lead agencies by sector, *108*
Critical Infrastructure Protection (PDD-63),
    31, 96–98, 107–9, 192
  budgets by department
    (1998–2000), *110*
  institutions established by, 109
  lead agencies by sector, *108*
  See also Counterterrorism
Critical Infrastructure Working Group
  (CIWG), 95
Crusade, 134
CT Security Group, 96
CTIO. *See* Counterterrorism information
  operations
CWB. *See* Collaborative white boarding
Cybercrimes, 44
Cyberspace Security, 99
Cyberterrorism, 44
Cyber war, 76
Cyberwar, Zapatistas and, 6–7

DASD S&IO organization, 31–32, *31*
Data networking, 15
DCI. *See* Director of Central Intelligence
DDOS. *See* Distributed denial of service
DEA. *See* Drug Enforcement Agency
Deception, 20
Decision superiority and subordinate
  concepts (Australia), *216*
Decisive Maneuver (Australia), 214
DefCons. *See* Defense Threat Conditions
Defence: Our Priorities, 215
Defence Science and Technology
  Organisation (DSTO), 213
*Defending America's Cyberspace, National
  Plan for Information Systems
  Protection Version 1.0, An
  Invitation to a Dialogue,* 109
Defense in-depth (DiD), 81–89
  DoDD 8500 instructions, 82–84
  IA DiD strategy, *84*
  operations, 89
  people, 82

technology, 82–89
See also Information protection
Defense Information Operations Council
    (DIOC), 46
Defense Information Systems Agency
    (DISA), 78
  IO architecture, 32, 33–35, *34*
Defense Intelligence Agency (DIA), 32,
    54, *55*
Defense Science Board, 24
Defense Science Board on Managing
    Information Dissemination
    (DSB-MID), 133
Defense Threat Conditions (DefCons), 79
Defensive information operations. *See*
    Information protection
Deny & deceive, Operation Iraqi
    Freedom, 156
Department of Commerce (DoC), 26, 27
  architecture, 41–43
  coordinating with Combatant
    Commander, 164
Department of Defense (DoD), 24, *25*–26
  CERT coordination, *81*
  combat support agencies, 32
  intelligence agencies, *55*
  Office of the Secretary of Defense and
    IO, 30–32
Department of Defense (DoDD) 8500
    instructions, 82–84
Department of Defense document (DoDD)
    TS3600.1, 22–23
Department of Energy (DoE), coordinating
    with Combatant Commander, 164
Department of Homeland Security (DHS),
    26, 27, 38, 44–45, 101
  coordinating with Combatant
    Commander, 164
  missions, 45
  organization of, *102*
  organizations under, 44
Department of Justice, IO architecture,
    43–44
Department of State (DoS), 24, *27*
  IO and, 40–41
  key members, 41
  traditional structure, 41
Destruction, 20
Detection, 68–69
DHS. *see* Department of Homeland Security
DIA. *See* Defense Intelligence Agency
DiD. *See* Defense in-depth
DIOC. *See* Defense Information Operations
    Council
Director of Central Intelligence (DCI),
    54, *55*
DISA. *See* Defense Information Systems
    Agency
Distributed denial of service (DDOS), 6–7
DNS Domain Name Security, 85

DoC. *See* Department of Commerce
*Doctrine for Intelligence Support to Joint
    Operations* (JP 2–0), 49–51
Domestic law, 183–85
Domestic terrorism, 91
  operations, 102–4
DoS. *See* Department of State
Downing, Wayne, 99
Drug Enforcement Agency (DEA),
    coordinating with Combatant
    Commander, 164
DSB-MID. *See* Defense Science Board on
    Managing Information
    Dissemination
DSTO. *See* Defence Science and Technology
    Organisation
Dunn, Myriam, *Information Age Conflicts:
    A Study of the Information
    Revolution and a Changing
    Operating Environment,* 14–15

EA. *See* Electronic attack
East Timor peace-enforcement mission. *See*
    Australian defence force
EC. *See* Electronic combat
EDT. *See* Electronic Disturbance Theater
Electronic attack (EA), 123–24, 146
Electronic combat (EC), 146
  Operation Enduring Freedom (OEF),
    150–52
  Operation Iraqi Freedom (OIF), 156–58
  See also Electronic warfare (EW);
    Information projection
Electronic warfare support (ES), 123–24
Electronic Disturbance Theater (EDT),
    6–7, 94
Electronic protection (EP), 123–24, 146
Electronic warfare (EW), 20
  electronic attack (EA), 123–24
  electronic protection (EP), 123–24
  electronic warfare support (ES), 123–24
  information operations and, 122–24
  See also Electronic combat (EC);
    Information projection
Eligible Receiver, 69
Eligible Receiver '9seven, 75–79, 232
Embedded journalists, 153
*Emergence of Noopoltik: Toward an
    American Information Strategy, The*
    (Arquilla and Ronfeldt), 9, 15
EP. *See* Electronic protection
ER '9seven. See Information protection;
    Eligible Receiver '9seven
ES. *See* Electronic warfare support
EuCom. *See* European Command
European Command (EuCom), 37
European Union (EU) and U.S. anti-
    terrorism summit (1998), 106–7
EW. *See* Electronic warfare
Extremist religious factions, 2

FALINTIL (Indonesia), 224
FEDCIRC. *See* Federal Computer Incident Response Cell
Federal Agency of Governmental Communication and Information (Russian special services) (FAPSI), 197
Federal Bureau of Investigation (FBI)
coordinating with Combatant Commander, 164
cyberterrorism and, 43–44
Federal Computer Incident Response Cell (FEDCIRC), 43
Firewalls, 86–87
1st Information Operations Command (1 IOC), 165–66
FIWC. *See* Fleet Information Warfare Center
Fleet Information Warfare Center (FIWC), 23, 166
Floodnet, 7
Foggy Bottom, 135
Foreign Affairs (Nye), 192–93
Fox, Vicente, 7
Franks, Tommy, 153, 159
Freedom, Patriot Acts and, 5
FRETLIN (Indonesia), 224
Friendly fire, 156–57

General Staff Academy (Russia), 200
Genocide, in Kosovo, 201
GIG. *See* Global Information Grid
Global area of responsibility, 15
Global Information Grid (GIG), 71
Global Network Operations Security Center (GNOSC), 33
Global Positioning System (GPS), 119–22, 157
Global War on Terror (GWOT), 101, 144, 159–61
GNOSC. *See* Global Network Operations Security Center
Goldwater-Nichols Act (1986), 209
GPS. *See* Global Positioning System
"Great Satan," 3
Guantanamo Bay, 154
Gulf War. *See* Operation Desert Storm
GWOT. *See* Global War on Terror

HARM anti-radiation missile, 157
Hart, Gary, 44
Haskell, Barbara, "Access to Society: A Neglected Dimension of Power," 11
HDR. *See* Humanitarian daily rations
Headquarters Australian Theatre (HQAST), 214
HFACs. *See* Human Factors Analysis Center
High-speed networking (ATM/100 Mbps Ethernet), 85
Hitler, Adolph, 4
Hoffman, Bruce, *Inside Terrorism,* 91

Homeland Security Council (HSC), 99
See also Department of Homeland Security
Honeypots and honeynets, 86, 88–89
HQAST. *See* Headquarters Australian Theatre
HSC. *See* Homeland Security Council
Huai Guomo, 207–8
Hughes, Karen, 134
Human factors analysis, 59–60
Human Factors Analysis Center (HFACs), 58
Human intelligence (HumInt), 52, 55, 59, 156
Humanitarian daily rations (HDR), 149
Humanitarian relief operations, in Afghanistan, 152
HumInt. *See* Human intelligence
Hussein, Saddam, 145, 153, 155, 158
Hyde, Henry, 136

IA Red Teams, 80–81
IA technical security measures, 85
IA Vulnerability Alert (IAVA), 80
IADS. *See* Integrated air defense system
IATAC. *See* Information Assurance Technology Analysis Center
IAVA. *See* IA Vulnerability Alert
IC. *See* Intelligence community
IDS. *See* Intrusion detection systems
IITF. *See* Information Infrastructure Task Force
ILOVEYOU virus, 39
Imagery intelligence (ImInt), 52, 55
IMI. *See* International military information
ImInt. *See* Imagery intelligence
In Athena's Camp (Arquilla and Ronfeldt), 11
Infinite Justice, 134
Influence operations
Operation Enduring Freedom (OEF), 147–50
Operation Iraqi Freedom (OIF), 152–56, 158
types, 146
InfoCon levels of IO attack, 79–80
Information
controlling, 231–32
defined, 1
views of, *10*
Information age, 163, 209
key features of, 9–10
operations in, 15
Information Age Conflicts: A Study of the Information Revolution and a Changing Operating Environment (Dunn), 14–15
Information assurance (IA), 71
See also Information protection
Information Assurance Technology Analysis Center (IATAC), 35, 36

Information attacks, 4
Information Infrastructure Task Force
        (IITF), 43
Information Operations (DoD JP 3–13), 36
Information Operations
        (DoDD S3600.1), 24
Information operations (IO)
    Operation Enduring Freedom (OEF),
        137–39
    origin of, 17
    targets of, *58*
Information operations (IO) campaigns.
        See Australian Defence Force; China;
        Kosovo; Russia
Information operations (IO) theory, 14–24
    capabilities and related activities, 20
    changing academic theory, 14–15
    command and control warfare (C2W),
        22–23, *22*
    and doctrine, 16–19
    evolution of the IO doctrine, 21–24
    information superiority components, *16*
    information superiority model, 17–18, *18*
    IO relationships across time, *17*
    operational concepts, 17
    vs. information warfare (IW), 19–21, *20*
Information operations legal issues, 178–85
    computer network attacks (CNAs), 115
    domestic law, 183–85
    law of armed conflict, 181–83
    Operation Enduring Freedom and, 142
    overview, 179–80
    peacetime treaties impact, 180–81
Information operations organizations,
        24–47, *27*
    architecture, 24–28, *25*, *27*
    combat support agencies, 32
    combatant commanders (CC) and, 36–39
    created under Clinton administration, 45
    Defense Information Systems Agency
        (DISA) architecture, 32, 33–35, *34*
    Department of Homeland Security, 44–45
    DoC IO architecture, 41–43
    DoD—Joint Chiefs of Staff (JCS) and,
        36–37
    DoD—Office of the Secretary of Defense
        and IO, 30–32
    DoJ IO architecture, 43–44
    DoS IO concerns, 40–41
    Information Assurance Technology
        Analysis Center (IATAC), 35, *36*
    intelligence community (IC), 39–40
    interagencies, 45–47
    interagency process and, 28–40
    National Communication systems (NCS),
        34, *35*
    National Security Agency (NSA) IO
        architecture, 32–33
    National Security Council (NSC)
        members and advisors, 28–29, *28*

    other presidential advisors, *30*
    partners, *25*
    top-level leadership, 26–28, *27*
    White House, 25–30, *27*
Information operations planning, 163–65
    agencies coordinating with Combatant
        Commander's staff, 164
    Joint Operations Planning and Execution
        System (JOPES), 173
    military IW service centers, 165–67
    operations plan, time-phased force
        deployment data and the IO cell,
        173–76, *175*
    rules of engagement (ROE) and, 185–86
    strategy-to-task planning, 168–71, *169–71*
    tools, 167–68
    tying together strategy-to-task planning
        and IO planning tools, 172
    as a warfighting strategy, 176–78
*Information Operations Staff Planning
        Manual* (Australian Defence Force
        Warfare Centre), 215
Information projection (offensive IO),
        111–62
    computer network attack (CNA),
        113–18, *114*
    electronic warfare (EW)
        electronic attack (EA), 123–24
        electronic protection (EP), 123–24
        electronic warfare support (ES), 123–24
        and IO, 122–24
    International Public Information (IPI)
        (PDD-68), 125
    IPI
        and Bush administration, 133–37
        and the Clinton administration, 125–27,
            129–31
        defined, 128
        history, 125–33
        key organizations, *130*
        and outside influences, 126–27
        success/failure of rate, 131–33
    offensive IO defined, 111–12
    offensive IO operations, 112–13
    offensive IO strategy, *114*
    Operation Enduring Freedom (OEF)
        coalition structure, 139–40
        electronic combat (EC) during, 150–52
        influence operations during, 147–50
        and IO, 137–39, 159–61
        network combat during, 152
        objective, 141–44
        Operation Iraqi Freedom (OIF)
            comparison, 144–62
        Venn diagram, *138*
    Operation Iraqi Freedom (OIF)
        electronic combat operations during,
            156–58
        influence operations during, 152–56
        IO summary, 159–61

network combat operations during, 158
Operation Enduring Freedom (OEF)
  comparison, 144–62
perception management, 124–44
space and its relationship with IO,
  118–22, *120*
Information protection (defensive IO),
  65–110
computer network attacks (CNAs), 75–79
  Eligible Receiver '9seven, 75–79
  Melissa virus, 77
  Moonlight Maze, 76–79
  Solar Sunrise, 75–76, 78–79
computer network defense (CND), 78–89
  biometrics, 87–88
  complex attack plan generators, 86
  defense in-depth (DiD), 81–89
  DoD CERT architecture, *78*
  DoD CERT coordination, *81*
  DoDD 8500 instructions, 82–83
  firewalls, 86–87
  honeypots and honeynets, 86, 88–89
  IA DiD strategy, *84*
  IA Red Teams, 80–81
  IA technical security measures, *85*
  IA Vulnerability Alert (IAVA)
    process, 80
  InfoCon levels, 79–80
  intrusion detection systems (IDS), 86–87
  J-6K protective layers, 84–85
  new trends, 85
  passwords, 87
  public key infrastructure (PKI), 85–86
  secure applications, 86
  smart cards, 86, 88
  SpaceCom June 2000 areas of
    concentration, 80
  virtual private network (VPN), 85–86
  virus scanners, 86
counterterrorism IO, 90–110
defensive IOs, 65–70
  defined, 65–66
  detection, 68–69
  process, 67–69, *67*
  protection, 68
  responding, 69
  restoration, 69
  risk assessment, *68*
information assurance
  growing threats to, *73*
  information environment protection,
    72–73, *73*
  information vulnerabilities, 72
  IO attack, detection and restoration, *74*
a view of defensive information
  operations, 89–90
  See also Counterterrorism
Information revolution, 11
Information Sharing and Analysis Centers
  (ISACs), 109

Information space, 192–94
Information superiority, *16,* 49, 59–60
  model, 17–18, *18*
Information Warfare Planning Capability
  (IWPC), 167–68
Information warfare (IW), 4
  courses, 8, 23
  military service centers, 165–67
  Russia and, 194–201
  vs. information operations, 19–21, *20*
Informational-psychological components
  of IO, Russia, 196–98
"Informationization of Russia on the
  Threshold of the twenty-first
  century," 197
Informatisionnaya voina (Information
  War), 196
Informatizatsiia, 192
Informatsionnoy bor'boy, 195
Infrastructure Protection Task Force
  (IPTF), 95
Inside Terrorism (Hoffman), 91
Insight, 135
Integrated air defense system (IADS), 157
Intelligence community, 54–60, *55*
  See also Intelligence support
Intelligence community (IC), 39–40
Intelligence cycle. *See* Intelligence support
Intelligence preparation of the battlespace
  (IPB), 113
  See also Intelligence support
Intelligence support, 49–63
  application of IO to, 50–51
  information superiority, *16,* 49
  intelligence community, 54–60
  agencies, *55*
intelligence cycle, 51–54
  phase 1: planning and directing, *52*
  phase 2: collection, *52*
  phase 3: processing and exploitation,
    52–53
  phase 4: analysis and production, *53*
  phase 5: dissemination and integration,
    53–54
  phase 6: evaluation and feedback, *54*
intelligence defined, 49–51
intelligence disciplines, *52*
intelligence products, *53*
intelligence superiority, 59
IO and joint intelligence preparation
  (IPB) for the battlespace, 56–60
  information effects to consider,
    57–58
  IO target examples, *58*
  step 1: defining the battlespace, 57
  step 2: describe battlespace effects,
    57–58
  step 3: adversary evaluation, 58–60
  step 4: determine adversary's course
    of action, 58–59

releaseability issues of IO, 60–63
Coalition Multi-Level Hexagon
Prototype (CHMP), 62–63
National Disclosure Policy (NDP-1),
61–62
INTERFET (Australia), 225–28
International cooperation, terrorism and,
105–7
International military information (IMI), 128
International public information (IPI), 40,
125–33
Kosovo and, 203
See also Information projection;
Perception management
International Public Information (PDD-68),
5, 40–41, 125–27, 129–34, 136
International terrorism, 92
Intrusion detection systems (IDS), 86–87
IPB. See Intelligence preparation of the
battlespace
IPI. See International public information
IPTF. See Infrastructure Protection Task
Force
Iraq. See Operation Desert Storm;
Operation Iraqi Freedom
ISACS. See Information Sharing and
Analysis Centers
IW. See Information warfare
IWPC. See Information Warfare Planning
Capability

JCIWS. See Joint Command, Control, and
Information Warfare School
JCS. See Joint Chiefs of Staff
JDAM. See Joint Direct Attack Munitions
JDISS. See Joint Deployable Intelligence
Support System
Jen Jui-Wen, 189
JFACC. See Joint Force Air Command
and Control
JFCom. See Joint Forces Command
JFHQ-IO. See Joint Force Headquarters
for IO
JFSC. See Joint Forces Staff College
JICs. See Joint Intelligence Centers
JIOAPP. See Joint Information Operations
Attack Planning Process
JIOC. See Joint Information Operations
Center
JIVA. See Joint intelligence virtual
architecture
Johnson, Lyndon, 119
Joint Chiefs of Staff (JCS), and IO, 36–37
Joint Command, Control, and Information
Warfare School (JCIWS), 8
Joint COMSEC Monitoring Activity
(JCMA), 33
Joint Deployable Intelligence Support
System (JDISS), 53, 55–56
Joint Direct Attack Munitions (JDAM), 157

Joint Doctrine for Electronic Warfare
(JP 3–51), 123–24
Joint Doctrine for Information Operations
(JP 3–13), 24, 111, 114, 122, 130,
165, 195, 215
definition of offensive IO, 111
Joint Force Air Command and Control
(JFACC), 117
Joint Force Headquarters for IO
(JFHQ-IO), 160
Joint Forces Command (JFCom), 38, 62, 112
Joint Forces Staff College (JFSC), 8
Joint Information Operations Attack
Planning Process (JIOAPP), 168, 172
Joint Information Operations Center (JIOC),
23, 166, 168
Joint Intelligence Centers (JICs), 55, 58
Joint intelligence preparation of the
battlespace. See Intelligence support
Joint intelligence virtual architecture
(JIVA), 54
Joint IO attack planning process, 170
Joint IO cell, 175
Joint IO doctrine, 65
Joint Operations Planning and Execution
System (JOPES), IO and, 173
Joint strategic capabilities plan (JSCP), 168
Joint Task Force-Computer Network
Operations (JTF-CNO), 33–34,
39, 79
Joint Vision 2010 (JV 2010), 17, 70–71, 209
Joint Vision 2020 (JV 2020), 18, 209, 232
Joint Warfare Analysis Center (JWAC), 167
Joint Worldwide Intelligence Communications
System (JWICS), 53, 55–56
JOPES. See Joint Operations Planning and
Execution System
JSCP. See Joint strategic capabilities plan
JTF-CNO. See Joint Task Force-Computer
Network Operations
JWAC. See Joint Warfare Analysis Center
JWICS. See Joint Worldwide Intelligence
Communications System

Karzai, Hamid, 149
KE. See Knowledge edge
Khobar Towers attack, 99
Knowledge edge (KE), 214, 216
Knowledge operations (KO), 216
KO. See Knowledge operations
Kokoshin, Andrei, 194
Kosovo, 41, 191–92, 199, 201–7
how an IO campaign may have succeeded,
205–7
IO after-action report, 204–5
news media and, 67
Operation Noble Anvil, 21, 46, 59
Operation Noble Anvil and IO use/misuse,
202–4
Kostin, N.A., 195

LAN. *See* Local area network
Land Information Warfare Activity, 23
Laquer, Walter, 91
Law
  domestic, 183–85
  See also Information operations legal issues
Laws of Armed Conflict (LOAC), 176, 181–83
"Left hook," 145
Legal issues. *See* Information operations legal issues
Liberalism, 14
Libiki, Martin, 211
Libya, air strikes against, 93
Lieberman, Joseph, 44
Lincoln Agreement, 221
LOAC. *See* Laws of Armed Conflict
Local area network (LAN), 115
Lopatin, Vladimar and V. D. Tsigankov
  Psikhotronnoe oruzhie i bezopastnost' rossii (Psychotronic Weapons and the Security of Russia), 196
  Psychotronic Weapons and the Security of Russia (Lopatin and Tsigankov), 197

Macro virus, 77
Managing Complex Contingency Operations (PDD-56), 125–26, 129–31, 133
MasInt. *See* Measurement and signature intelligence
Mattis, James M., 153
Measurement and signature intelligence (MasInt), 52, 151
Measures of effectiveness (MOE), 112, 154
Media
  cooperation with military in national emergencies, 5
  Kosovo and, 203
  Zapatista movement and, 6–7
Melissa virus, 77
Middle East policy, 4
Military, cooperation with media in national emergencies, 5
Military information support team (MIST), 222–23
Military IW service centers, 164–67
Military operations other than war (MOOTW), 46, 181
Milosevic, Slobodan, 203–204
MIST. *See* Military information support team
MOE. *See* Measures of effectiveness
Montville, Joseph, 104
Moonlight Maze, 76–79
  See also Information protection
MOOTW. *See* Military operations other than war

Morganthau, Hans, *Politics Among Nations: The Struggle for Power and Peace*, 10
Musharraf, Pervez, 139
Myers, Richard B., 160

NAIP. *See* National Infrastructure Assurance Plan
Nation-states, 14
National Capital Region (2000), 104
National Command Authorities. *See* SecDef
National Commission on Terrorism Act (1998), 106
National Communications Systems (NCS), 34–35
  members, 34, 35
National Defense University (NDU)
  Joint Forces Staff College, 23
  Joint Information Warfare Staff and Operations Course, 23
  School of Information Warfare and Strategy, 23
National Director for Combating Terrorism, 99
National Disclosure Policy (NDP-1), 61–62
National Information Assurance Partnership (NIAP), 43
National Information Infrastructure (NII), 43
National Information Systems Weekly Incident Summary, 33
National Infrastructure Assurance Council (NIAC), 98, 100, 109
National Infrastructure Assurance Plan (NAIP), 98
National Infrastructure Protection Center (NIPC), 44, 95, 109
National Institute of Standards and Technology (NIST), 42–43
  current areas of interest, 43
  task of, 43
National Military Joint Intelligence Center (NMJIC), 55
National Plan for Information Systems Protection, 98
National Reconnaissance Office (NRO), 55
National Security Act (1947), 54
National Security Agency (NSA), IO architecture, 32–33
National Security Council (NSC), 27, 28–29, 99
  members and advisors, 28
National Security Incident Response Center (NSIRC), 33
National security strategy (NSS), 3, 163
National Security Telecommunications and Information Systems Security Council (NSTISSC), 33
National Strategy for Combating Terrorism, 101

National Strategy for Homeland Security, 100–101
National Strategy for the Physical Protection of Critical Infrastructures and Key Assets, 101
National Strategy to Secure Cyberspace, 101
National Technological Information Association, 43
National Telecommunications and Information Administration (NTIA), 42–43
National-separatist terrorism, 93
National War College, 23
NATO Civil Military Cooperation (CIMIC), 205–6
NATO, Kosovo and, 191–92, 199, 201–7
NC. *See* Network combat
NDU. *See* National Defense University
Neoliberals, 11
Neo-realism, 14
Neorealists, 11
Network combat (NC), 146
    Operation Enduring Freedom (OEF), 152
    Operation Iraqi Freedom (OIF), 158
New religions terrorism, 94
New York Times, 153
NGOs. *See* Nongovernmental organizations
NIAC. *See* National Infrastructure Assurance Council
NIAP. *See* National Information Assurance Partnership
NII. *See* National Information Infrastructure
NIPC. *See* National Infrastructure Protection Center
NIST. *See* National Institute of Standards and Technology
Nius Blong Peace (Peace News), 223
NMJIC. *See* National Military Joint Intelligence Center
Non-governmental organizations (NGOs), 2, 6, 46, 94, 104–5, 107, 129
    combatant commanders and, 46–47
NorthCom. *See* Northern Command
Northern Command (NorthCom), 38
NRO. *See* National Reconnaissance Office
NSA. *See* National Security Agency
NSC. *See* National Security Council
NSIRC. *See* National Security Incident Response Center
NSS. *See* National security strategy
NSTISSC. *See* National Security Telecommunications and Information Systems Security Council
NTIA. *See* National Telecommunications and Information Administration
Nye, Joseph S.
    Bound to Lead, 11
    Foreign Affairs, 192–93

Observe, orient, decide and act (OODA), 214
Offensive information operations. *See* Information projection
Office of Global Communications (OGC), 135
Office of Homeland Security, 96, 99–100
Office of the National Coordinator for Security, Infrastructure Protection and Counter-Terrorism, 96–97
Office of the Secretary of Defense (OSD), 30–32
Office of Strategic Influence (OSI), 135–36, 159–60
Office of Terrorism Preparedness, 98
OGC. *See* Office of Global Communications
Oklahoma City bombing (1995), 4, 91, 107
Omar, Mohammed, interview with Voice of America, 5
Ona, Francis, 221
OODA. *See* Observe, orient, decide and act
Open-source intelligence (OsInt), 52
Operation BELISI (Bougainville), 219–24, 228–29
Operation Desert Storm (Iraq), 14, 16, 21, 121, 125, 128, 137, 144, 145, 156
    Australia and, 213
    compared to Operation Enduring Freedom, 142
Operation Enduring Freedom (OEF), 4, 21, 67, 93, 107, 133–34, 137–52
    coalition structure (alliances), 139–40
    compared to Operation Desert Storm (ODS), 142
    electronic combat operations during, 150–52
    influence operations during, 147–50
    information operations (IO) and, 137–39
    network combat operations during, 152
    objective of, 141–44
    Operation Iraqi Freedom (OIF) comparison, 144–47
    summary, 159–61
    Venn diagram, *138*
    See also Information projection
Operation Iraqi Freedom (OIF), 1, 4, 21, 144–45
    electronic combat operations during, 156–58
    influence operations during, 152–56
    network combat during, 158
    summary, 159–61
    See also Information projection
Operation LAGOON (Bougainville), 220
Operation Noble Anvil (Kosovo), 21, 46, 59, 201–7
Operation Northern Watch (Iraq), 147

Operation Rausim Kwik (Bougainville), 221
Operation Southern Watch (Iraq), 147
Operation STABILISE (East Timor),
    224–29
Operation Uphold Democracy (Haiti), 154
Operations plan (OPLAN), 173
Operations security (OpSec), 20, 66,
    152–54, 158
OPLAN. See Operations plan
OpSec. See Operations security
OSD. See office of the Secretary of Defense
OSI. See Office of Strategic Influence
OsInt. See Open-source intelligence
Owens, William, 209

PacCom. See Pacific Command
Pacific Command (PacCom), 38
Panarin, Igor, 197
Papua New Guinea. See Australian Defence
    Force
Passwords, 87
Patriot Acts, 5, 161
PCC. See Policy Coordinating Committee
PCCIP. See Presidential Commission on
    Critical Infrastructure Protection
PCIPB. See President's Critical
    Infrastructure Protection Board
Peace Monitoring Group (PMG), 219, 223
Peacetime treaties, impacting IO, 180–81
Pentagon, September 11 terrorist attack,
    3, 144
People's Republic of China. See China
Perception management, 124–44
    See also Information projection;
        International public information;
        Psychological operations
Perl, Raphael, 105–6
PKI. See Public key infrastructure
Planning. See Information operations
    planning
PMG. See Peace Monitoring Group
POG. See Psychological Operations Group
Policy Coordinating Committee (PCC), 134
Politics Among Nations: The Struggle for
    Power and Peace (Morganthau), 10
Post-cold war era, 2
Powell, Colin, 50–51, 143
Power, 231–32
    actual levels of, 12
    changing, 11, 13–14
    in cold war era, 13
    defined, 10–14
    information and, 13
    military and asymmetric threats, 14
    theoretical levels of, 12
    views of, 9, 10
Practices for Securing Critical Information
    Assets (Critical Infrastructure
    Assurance Office), 109

President, 26, 27
Presidential advisors, 30
Presidential Commission on Critical
    Infrastructure Protection (PCCIP),
    95–96, 108
President's Critical Infrastructure Protection
    Board (PCIPB), 99
Principals Committee, 26
Private voluntary organizations (PVOs), 46
Project Takari (Australia), 213
Propaganda, terrorism and, 4–5
Prophet (unmanned aerial vehicle), 151
Protection. See Information protection
Protivoborstvom, 195
Psikhotronnoe oruzhie i bezopastnost' rossii
    (Psychotronic Weapons and the
    Security of Russia) (Lopatin and
    Tsigankov), 196
Psychological Operations Group (POG), 173
Psychological operations (PSYOPS), 4, 20,
    128, 131, 135, 145–47, 159–60,
    172–74, 177
    in Afghanistan, 147–51
    in Iraq, 152–56, 158
    in Kosovo, 203, 205–7
    See also Information projection;
        International public information;
        Perception management
Psychotronic Weapons and the Security
    of Russia (Lopatin and
    Tsigankov), 197
"Psycho virus," 196
PSYOPS. See Psychological operations
Public affairs (PA), 20
Public key infrastructure (PKI), 85–86
Public opinion, 231
Putin, Vladimir, 139
PVOs. See Private voluntary organizations

RAAF. See Royal Australian Air Force
RAN. See Royal Australian Navy
Raskols, 220
Realism, 14
    structural, 15
Red Star, 199
Red Team, 75
Reflexive control, 197–98
Relevant information, 49
Religious fundamentalist terrorism, 94
Rendon Group, 134
Rendon, John, 134
Renuart, Victor, 157
Republican Guard (Iraq), 145
Responding, 69
Restoration, 69
Ridge, Tom, 99
Right wing terrorism, 93
Rio Treaty by the Organization of American
    States, 143

Rise of the Virtual State, The
    (Rosecrance), 11
Risk assessment process, *68*
Rivera, Geraldo, 153
Rodriguez, Mark, 135
ROE. *See* Rules of engagement (ROE)
Rogue states, 2
Rosecrance, Richard, *Rise of the Virtual
    State, The,* 11
Rot I Go Long Peace (The Road to
    Peace), 223
Royal Australian Air Force (RAAF), 218
Royal Australian Navy (RAN), 218
Ruby Ridge, 107
Rudman, Warren, 44–45
Rules of engagement (ROE) planning, 179,
    185–86
Rumsfeld, Donald, 135, 152–53
Russia, 189–201
    computer network attacks and, 115
    defense budget, 2
    information space, 192–94
    information-specific interests, 190–91
    information superiority, 191–92
    information warfare (IW) terminology
        and theory, 194–201
    informational-psychological
        components of IO, 196–98
    military-technical components of IO,
        198–200
    systemological aspects, 200–201
Russian, *Moonlight Maze* event, 76
Russian State Technical Commission, 192

SAPs. *See* Special access programs
Schwarzkopf, Norman, 145
SCUD missile launchers, 156
SEAD. *See* Suppression on enemy air defense
SecDef (National Command Authorities),
    26, *27, 36*
Secretary of defense, 26
"Secret United States Only No
    Foreigners," 60
Secret Weapons of Information Warfare, 197
Secretary of state, 26
Secure applications, 86
Security policy coordination, 32
Seneca, 163
September 11 terrorist attacks, xvii-xviii,
    2–6, 91, 232
    arrests after, 142
    background, 144
    changes in government's approach after,
        99–102, *102,* 133–37
    lessons from, 4
    post-September 11th CTIO
        organizational structure, *102*
    pre-September 11th CTIO organizational
        structure, *100*
    psychological impact of, 160–61

Sergeyev, Igor, 198–99
SigInt. *See* Signals intelligence
Signals intelligence (SigInt), 32, 52, 55, 151
SIO. *See* Special information operations
Smart cards, 86, 88
Smith-Mundt Act (1948), 132
Social revolutionary left terrorism, 93
SOCom. *See* Special Operations Command
SOFA. *See* Status of forces agreement
Soft power, 11
Solar Sunrise, 75–79
    See also Information protection
Somalia, 16
    U.S. Army Rangers dragged through
        streets of Mogadishu, 66, 94
SouthCom. *See* Southern Command
Southern Command (SouthCom), 38
Soviet Union, 2
    Afghanistan and, 14
    fall of, 1
    military force (1975–1985), 21
Space
    commercialization of, 119, 121–22
    information operations and, 118–22, *120*
    surveillance, 119, *120*
    See also Information projection
SpaceCom (U.S. Space Command), 33–34,
    38–39, 80
Special access programs (SAPs), 60
Special information operations (SIO), 218
Special Operations Command (SOCom), 38
Special technical operations (STO), 60
Specter, Arlan, 44
State Duma's Security Committee, 196
State supported terrorism, 93
Status of forces agreement (SOFA), 181
STO. *See* Special technical operations
StratCom. *See* Strategic Command
Strategic Command (StratCom), 38, 79
Strategy-to-task planning, 168–72,
    169–71
Structural realism, 15
Subcommandante Marcos, 6–7
Sudan, 93
Sun Tzu, 49, 51
Suppression on enemy air defense (SEAD),
    146, 156
Sussman, Michael A., "Critical Challenges
    from International High-Tech and
    Computer-Related Crime at the
    Millennium," 184
Swarming attacks, 9

Tactical air-launched decoy (TALD), 157
TALD. *See* Tactical air-launched decoy
Taliban, 141–42, 147
    blocked financing, 143
    Mullah Mohammed Omar, 5
    propaganda pamphlets, 149–50
    Voice of Sharia radio, 150

Tarawa Peace Conference, 220, 222
TechInt. *See* Technical intelligence
Technical intelligence (TechInt), 52
Telecommunications Act (1996), 43
TEP. *See* Theater engagement plan
Terrorism
  combating. *See* Counterterrorism
  defined, 91–92
  incidents in the U.S. (1993–1999), 97
  propaganda and, 4–5
  psychological impact of, 160–61
  types of, 93–94
  U.S. policy tenets, 93
  See also Counterterrorism; Information projection; Information protection
Terrorism Preparedness Act (2000), 98
Terrorist groups, 2
Theater engagement plan (TEP), 41
Theater Engagement Planning (CJCSM-3113.1), 164
Theater security cooperation plan (TSCP), 112
Time-phased force deployment data (TPFDD), 173–76, *175*
Time zones, 9
TMG. *See* Truth Monitoring Group
Top-level leadership, 26–28, *27*
TOPOFF, 104
"Top Secret/Sensitive Compartmented Information," 60
TPFDD. *See* Time-phased force deployment data
Track Two Diplomacy, 104–7
TransCom. *See* Transportation Command
Transportation Command (TransCom), 38
Treaties, peacetime, 180–81
Truth Monitoring Group (TMG), 219
TSCP. *See* Theater security cooperation plan
Tsymbal, V. I., 194
2000 Defence White Paper, 216–17

UAV. *See* Unmanned aerial vehicle
United Nations (UN)
  Clinton's (1998) speech to, 106
  international legal environment and, 179–80
United Nations (UN) Security Council, Iraq invasion and, 143, 146
United States government (USG), 4
  information operations (IO) organizations, 27
United States Information Agency (USIA), 40
Universal Declaration of Human Rights, 106
Unmanned aerial vehicle (UAV), 151
U.S. embassies, attacks on, 99, 106
U.S. and European Union (EU) anti-terrorism summit (1998), 106–7

U.S. *Policy on Counter-Terrorism* (PDD-39), 94–95
U.S. Space Command. *See* SpaceCom
USG. *See* United States government
USIA. *See* United States Information Agency
USS Cole attack, 99

Venn diagram, Operation Enduring Freedom (OEF), *138*
Vice-president, 26
Vietnam war, 14
Virtual private network (VPN), 85–86
Virus scanners, 86
Viruses
  ILOVEYOU, 77
  Melissa, 77
  See also Information protection
VOA. *See* Voice of America
Voice of America (VOA), interview with Mullah Mohammed Omar, 5
Voice of Sharia radio, 150
"Voice of Youth," 154
VPN. *See* Virtual private network
Vulnerabilities, 71–72, *72*
  See also Information protection

Waco, 107
Waller, J. Michael, 159, 160
Warfighting strategy, IO as, 176–78
War on terrorism, characteristics of, 141
Warrior Flag, 69
Warsaw Pact, 2
Washington Monument, *120*
Washington Post, 135
Washington Times
  IPI and, 127, 131–32
  "Professor Albright Goes Live," 131–32
Weapons of mass destruction (WMD), 94, 156
White House, 25–30, *27*
Wilhelm, Charles, 112
Wireless technology, 85
WMD Preparedness Group, 96, 97
WMD. *See* Weapons of mass destruction
Wolf, Frank, 106
Wolfowitz, Paul, 154
World Trade Center
  1993 bombing, 99, 107
  September 11 terrorist attacks, 3, 90, 144
Worms, 77

Xu Chuangjie, 209–10

Y2K, 29–30, 35
Y2K Task Force, 43

Zapatistas, 6–7
Zedillo, 6